NSTA Tool Kit
for Teaching
EVOLUTION

NSTA Tool Kit
for Teaching
EVOLUTION

Judy Elgin Jensen

National Science Teachers Association

Arlington, Virginia

How paramount the future is to the present
when one is surrounded by children.
—Charles Darwin, in a letter to W.D. Fox, March 5, 1852

Claire Reinburg, Director
Jennifer Horak, Managing Editor
Judy Cusick, Senior Editor
Andrew Cocke, Associate Editor
Betty Smith, Associate Editor

ART AND DESIGN
Will Thomas, Jr., Director
Joe Butera, Graphic Designer

PRINTING AND PRODUCTION
Catherine Lorrain, Director
Nguyet Tran, Assistant Production Manager

NATIONAL SCIENCE TEACHERS ASSOCIATION
Francis Q. Eberle, Executive Director
David Beacom, Publisher

Copyright © 2008 by the National Science Teachers Association.
All rights reserved. Printed in the United States of America.
11 10 09 08   4 3 2 1

LIBRARY OF CONGRESS CATALOGING-IN-PUBLICATION DATA
Jensen, Judy Elgin.
  NSTA tool kit for teaching evolution / National Science Teachers Association ; Judy Elgin Jensen.
       p. cm.
  Includes bibliographical references.
  ISBN 978-1-933531-46-5
  1. Evolution (Biology)—Study and teaching. I. National Science Teachers Association, II. Title.
  QH362.J46 2008
  576.8071—dc22
                    2008037810

NSTA is committed to publishing material that promotes the best in inquiry-based science education. However, conditions of actual use may vary, and the safety procedures and practices described in this book are intended to serve only as a guide. Additional precautionary measures may be required. NSTA and the authors do not warrant or represent that the procedures and practices in this book meet any safety code or standard of federal, state, or local regulations. NSTA and the authors disclaim any liability for personal injury or damage to property arising out of or relating to the use of this book, including any of the recommendations, instructions, or materials contained therein.

PERMISSIONS
You may photocopy, print, or email up to five copies of an NSTA book chapter for personal use only; this does not include display or promotional use. Elementary, middle, and high school teachers *only* may reproduce a single NSTA book chapter for classroom- or noncommercial, professional-development use only. For permission to photocopy or use material electronically from this NSTA Press book, please contact the Copyright Clearance Center (CCC) (*www.copyright.com*; 978-750-8400). Please access *www.nsta.org/permissions* for further information about NSTA's rights and permissions policies.

Featuring sciLINKS®—a new way of connecting text and the Internet. Up-to-the minute online content, classroom ideas, and other materials are just a click away. For more information go to www.scilinks.org/faq/moreinformation.asp.

Table of Contents

Acknowledgments..**viii**

Introduction..**ix**
    What's in the *NSTA Tool Kit for Teaching Evolution*?
    Supporting Your Efforts

Chapter 1: The History and Science of Evolution...**1**
    The Nature of Science and Scientific Thinking
    What Is Evolution?
    History of Evolutionary Thinking
    Darwin's Thought Process
    The Modern Synthesis
    The Genetics of Evolution

Chapter 2: Active Evolution Instruction............**17**
    Showing Evolutionary Relationships With Cladograms
    Natural Selection and Antibiotic Resistance
    Using Ring Species to Demonstrate Human Evolution
    More Resources

Chapter 3: Social Challenges of Promoting Evolution-Only Instruction............................**45**
    What's *Wrong* With "Teaching the Controversy"
    Evolution Instruction in the Courtroom
    Evolution Instruction in the Classroom
    Leading the Way for Evolution Instruction

Chapter 4: What *You* Can Do..........................**63**
    Community Awareness
    The NSTA Community
    Need More Help?

References..**67**

Index..**69**

Acknowledgments

I would like to acknowledge with gratitude the contributions by staff from the National Center for Science Education, specifically Louise S. Mead, PhD, and Eugenie C. Scott, PhD. Louise and Eugenie's content knowledge and the background they could share about the underlying social issues influencing the teaching of evolution were indispensable.

I would also like to thank Mary Bigelow of the SciLinks team for her help in vetting and assembling great online and print resources related to the teaching of evolution. Mary also reviewed an early draft of the text, offering helpful comments and encouragement. Finally, I would like to thank Rodger Bybee, PhD, and Juliana Texley, PhD, for their insightful reviews of the draft manuscript.

Introduction

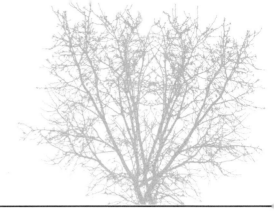

You are midway through your district curriculum, and you need to pull together the subjects you have been teaching so far—nature of science, cells, biochemistry, and genetics. Have you considered teaching evolution as a way of tying these topics together? That's right, *evolution*. Or are you one of the many biology or life science teachers who say, *"I don't teach evolution because ..."*

1. "I don't want to be in the middle of this controversial topic. I don't know how to best answer parents, students, or others who ask why I don't teach alternative theories."
2. "I want students to develop their own opinions, yet it's difficult to frame evolution instruction in a way that leaves students' minds open but doesn't equivocate."
3. "I don't have the class time to spend dealing with parent and student biases against evolution instruction."
4. "I need more content support to feel confident about my instruction, or even to supply simple answers to basic questions about evolution and the nature of science."
5. "I need hands-on, visual teaching aids."
6. "I don't spend time on a potentially controversial topic that's not assessed on my state-mandated test."
7. "I'm unsure what my district's policy is on teaching evolution and I'm concerned that I won't receive total support from the administration."
8. "I don't really know what to say about 'intelligent design' if the topic comes up."
9. "I disagree with the policy that allows students who are offended to skip class and the negative message this sends to the rest of my class."
10. "I have to either steer clear of human evolution or 'teach both sides.'" (Omenn 2006)

If you shy away from teaching evolution, you are not alone. The 10 statements just listed are adapted from an American Association for the Advancement of Science (AAAS) survey of teachers who identified and ranked these challenges.

Introduction

Topic:
Evolution
teaching
resources

Go to:
www.SciLinks.org

Code: ETK001

To help you overcome these challenges and teach evolution successfully, the National Science Teachers Association (NSTA) has partnered with the National Center for Science Education (NCSE) on the *NSTA Tool Kit for Teaching Evolution*. You can also use this guide to enhance your teaching of the National Science Education Standards listed in Table 1 as well as those associated with evolution in your state and district curricula.

Table 1. National Science Education Standards Pertaining to the Teaching of Evolution

Grades 5–8 Life Science

Topic: Regulation and behavior (p. 157)
- How a species moves, obtains food, reproduces, and responds to danger is based in the species' evolutionary history.
- All organisms must be able to obtain and use resources, grow, reproduce, and maintain stable internal conditions while living in a constantly changing external environment.
- An organism's behavior evolves through adaptation to its environment.

Topic: Diversity and adaptations of organisms (p. 158)
- Biological evolution accounts for the diversity of species developed through gradual processes over many generations.
- Species acquire many of their unique characteristics through biological adaptation, which involves the selection of naturally occurring variations in populations.
- Biological adaptations include changes in structures, behaviors, or physiology that enhance survival and reproductive success in a particular environment.

Topic: Structure and function in living systems (p. 156)
- Living systems at all levels of organization demonstrate the complementary nature of structure and function.

Topic: Reproduction and heredity (p. 157)
- The characteristics of an organism can be described in terms of a combination of traits. Some traits are inherited and others result from interactions with the environment.

Grades 9–12 Earth Science

Topic: The origin and evolution of the Earth system (p. 190)
- The evolution of life caused dramatic changes in the composition of the Earth's atmosphere, which did not originally contain oxygen.

Grades 9–12 Life Science

Topic: Biological evolution (p. 185)
- Evolution is the consequence of the interactions of the potential for a species to increase its numbers.
- Evolution is the consequence of the interactions of the genetic variability of offspring due to mutation and recombination of genes.
- Evolution is the consequence of the interactions of a finite supply of the resources required for life.
- Evolution is the consequence of the interactions of the ensuing selection by the environment of those offspring better able to survive and leave offspring.
- The great diversity of organisms is the result of more than 3.5 billion years of evolution that has filled every available niche with life-forms.
- Natural selection and its evolutionary consequences provide a scientific explanation for the fossil record of ancient life-forms, as well as for the striking molecular similarities observed among the diverse species of living organisms.
- Organisms are classified into a hierarchy of groups and subgroups based on similarities, which reflect their evolutionary relationships.

Topic: The behavior of organisms (p. 187)
- Behaviors often have an adaptive logic when viewed in terms of evolutionary principles.

Source: National Research Council. 1996. *National Science Education Standards*. Washington, DC: National Academy Press.

Here's more data on your peers in the classroom. In a scientific survey of practicing biology teachers, 98% taught evolution and 78% taught human evolution. The average amount of time spent on evolution—13

Introduction

hours—is not adequate for students to understand the importance of this topic. Notably, teachers with more science background (more university-level science courses) taught more evolution (Berkman, Pacheco, and Plutzer 2008). The *NSTA Tool Kit for Teaching Evolution* will help you integrate evolution into your courses so that you can successfully teach it more often, regardless of your prior science experience.

What's in the *NSTA Tool Kit for Teaching Evolution?*

This go-to reference guide includes information and activities that you can use in class tomorrow. Organized into four distinct chapters, this book will

- refresh your background knowledge and help you know what to emphasize in your evolution instruction,
- support you in hands-on instruction,
- clarify the controversy to help you take advantage of the opportunities in evolution instruction, and
- identify organizations that support your professionalism in this endeavor.

As you read the *Tool Kit*, log on to *www.scilinks.org* to access the SciLinks listed throughout the book for more content, activities, and assessments. The SciLinks program was developed by the National Science Teachers Association (NSTA) to supplement classroom discussions of key science topics. This searchable database automatically filters, centralizes, and saves helpful science websites found by NSTA experts. The sites are organized by topic and presented in an accessible format for teachers and students. Use them during lectures or group activities or as homework assignments. Full access to all SciLinks tools and topics requires registration, which is free and easy. Once you are registered, provide your students with your ID # and let them explore on their own.

Q&A

Throughout this book you will also find abridgements (like the following) of a question-and-answer (Q&A) document written by past NSTA president Gerald Skoog to help dispel misconceptions about evolution. The complete text of the Q&A document is online at *www.nsta.org/publications/evolution.aspx#qanda*.

Q: *Why have challenges to the teaching of evolution increased so dramatically in recent years?*

A: Throughout the 20th century, special-interest groups have worked to prohibit, deemphasize, or neutralize the teaching of evolution in the nation's public schools. Recently, cultural, religious, judicial, legal, and other factors have shaped the nature, intensity, and success of these efforts and challenges. The generally strong presence of biological evolution in the national and state science standards and science textbooks has catalyzed the actions of special-interest groups who currently have considerable political influence and question the legitimacy of biological evolution and its place in the science curriculum and public thought.

Q: *What do anti-evolution groups want?*

A: The answer is complicated because many groups hold different beliefs about the history and nature of life on Earth. Groups on one end of the spectrum seek to replace or "balance" the teaching of evolution with a literal biblical interpretation of creation, while others, to sidestep the overtly religious argument, call for teaching the so-called "weaknesses" of evolution. Still others believe that a myriad of ideas, scientifically supported or not, should be taught out of "fairness."

Q: *I'm frustrated at the amount of time and attention being devoted to the evolution issue, especially when I have so many other demands and challenges in the classroom. Shouldn't we all just keep a low profile and hope the issue goes away?*

A: We understand your frustrations. We've heard from many of you that the dialogue on this issue is causing undue stress and usurping valuable time. Now more than ever, we need the voice of the nation's science teachers to be heard. The stakes are simply too high. We recommend that you use this opportunity to educate and inform students, school leaders, and community members about the nature of science and what it can and can't tell us about the world. Rest assured you are not alone in this effort. NSTA stands ready to support you in any way and is working at the national level to keep evolution—and sound science—in its rightful place in the science curriculum.

Introduction

Research shows . . .

Three research studies offer insight into the controversy surrounding the teaching of evolution. Throughout the book statistics from each of these studies are boxed off and cited under the heading "Research shows . . ."

The first study—a 2006 survey of 1,000 randomly selected individuals conducted by the Coalition of Scientific Societies—focuses on the attitudes of Americans (the general public, not just teachers) regarding science education and the teaching of evolution, creationism, and intelligent design. Attitudes vary depending on the education level, religious background, politics, marital status, race, sex, and size of community of the respondents. Knowing the demographics of your school's community can help you better frame your stance of teaching evolution as a valid scientific explanation.

The second study is by three political science experts at The Pennsylvania State University who surveyed teachers about the teaching of evolution in the 2006–2007 academic year. The researchers found that personal beliefs and college-level courses in evolutionary biology impact teaching practices.

A third study was conducted in 2005 by the Pew Forum on Religion and Public Life, which charges itself with promoting a deeper understanding of issues at the intersection of religion and public affairs. This telephone survey reached a nationwide sample of 2,000 adults.

Research shows...

- Even though nearly half of Americans believe that humans evolved over time, only 26% say they favor teaching evolution only in the public schools while 64% favor teaching creationism along with evolution. Another aspect of the survey revealed that 38% would teach creationism instead of evolution although 49% opposes teaching creationism only. "These findings strongly suggest that much of the public believes it is desirable to offer more viewpoints where controversial subjects in the schools are concerned" (Pew Forum on Religion and Public Life 2005, p. 10).

- In the classroom, 69% of biology teachers spend from 3–5 hours on general evolutionary processes. In addition, about 25% of biology teachers spend at least 1–2 hours on creationism. While some of these teachers address creationism in response to student inquiries or to criticize it, nearly half of that 25% say that creationism is a valid scientific alternative to Darwinian explanations for the origin of species (Berkman, Pacheco, and Plutzer 2008, Table S2).

Supporting Your Efforts

NSTA supports your efforts to teach evolution as a major unifying concept in your biology or life science class, as outlined in its position statement at *www.nsta.org/about/positions/evolution.aspx*. Local curriculum and attitudes can make this endeavor difficult, but we hope the *NSTA Tool Kit for Teaching Evolution* supports you in your current practices and provides new ideas and resources that add depth to your instruction.

Chapter 1: The History and Science of Evolution

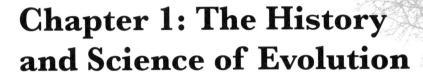

As discussed in the Introduction, the more exposure teachers have to evolutionary biology, the more likely they are to teach the concept of evolution in their classrooms and the more confident they feel in their instruction. This chapter cannot replace a college-level evolutionary biology course; however, the information provided here, along with your own research, can extend or fill gaps in your background knowledge of the history and science of evolution. For additional information, visit the referenced internet sites, which offer greater depth or different approaches to the particular topics.

The Nature of Science and Scientific Thinking

Science is a method of explaining the natural world and assumes that anything that can be directly or indirectly observed or measured is amenable to scientific investigation. Science also assumes that the universe operates according to regularities that can be discovered and understood through scientific investigations. The testing of various explanations of natural phenomena for their consistency with empirical data is an essential part of the methodology of science. Explanations that are not consistent with empirical evidence or cannot be tested empirically are not a part of science. As a result, explanations of natural phenomena based on evidence are scientific explanations. Those explanations based on personal beliefs, religious values, and superstitions are not scientific.

The most important scientific explanations are called theories. In ordinary speech, theory is often used to mean guess or hunch, whereas in scientific terminology, a theory is a set of universal statements that explain some aspect of the natural world, such as the big bang theory or the theory of evolution. Theories may include scientific laws as part of the explanation. A law differs from a theory in that a law describes an observed pattern in nature without attempting to explain it, such as the law of independent assortment of alleles or the law of gravity.

Theories are powerful tools. Scientists seek to develop theories that

- are firmly grounded in and based on evidence,
- are logically consistent with other well-established principles,
- explain more than rival theories, and
- have the potential to lead to new knowledge.

The History and Science of Evolution

The body of scientific knowledge changes as new observations and discoveries are made. Theories and other explanations change. New theories emerge; others are modified or discarded. Throughout this process, theories are formulated and tested on the basis of evidence, internal consistency, and their explanatory power.

Q: *How do I respond to the "evolution is just a theory" argument?*

A: A theory is a well-substantiated explanation of some aspect of the natural world that can incorporate facts, laws, inferences, and tested hypotheses, while a fact is an observation that has been repeatedly confirmed (NAS 1998). Thus, a scientific theory is not just a hunch or a guess. Scientific theories continue to change as new observations and discoveries are made. Based on research, testing, and observation, the theory of evolution is the best scientific explanation we have for how life on Earth has changed and continues to change.

Q: *Some say that evolution cannot be proven because we were not there to see it happen. How do I respond?*

A: Science allows us to study past events and life even though we were not present to observe them. We know, for example, that ancient cultures existed because we can study artifacts that have been left behind, and we know where bodies of water used to exist based on the information we find in the layers of soil and rock, as is currently being done by unmanned vehicles on Mars. In addition, scientists are using cosmic microwaves to develop an understanding of the beginning seconds of the universe's existence. Scientists use the data from these microwaves to develop basic parameters that characterize the universe, including its age, geometry, composition, and weight.

As these examples show, data derived by different means are used as evidence to support scientific theories. It is also important to recognize that scientific theories and laws are subject to change as a result of new evidence. Thus, the goal of science is not to prove, but to explain.

What Is Evolution?

Biological evolution refers to descent with modification, the scientific theory that states that living things have diverged from shared ancestors.

In the broadest sense, evolution can be defined as the idea that the universe has a history, that change through time has taken place. If we look at the galaxies, stars, the planet Earth, and life on Earth, we see that things today are different from what they were in the past and that galaxies, stars, planets, and life-forms have evolved. Abundant and consistent evidence from astronomy, physics, biochemistry, geochronology, geology, biology, anthropology, and other sciences shows that evolution has taken place.

> **Research shows...**
> - In the classroom, 60% of biology teachers agree or strongly agree that evolution serves as the unifying theme for the content of the biology course. Yet, 36% of respondents disagree or strongly disagree with this statement (Berkman, Pacheco, and Plutzer 2008, Table S2).

There is no debate among scientists about whether evolution has taken place. The latest edition of *Science, Evolution, and Creationism*, released by the National Academy of Sciences and the Institute of Medicine (2008), states that evolution is a core concept in biology based both on the study of past life-forms and on the study of the relatedness and diversity of present-day organisms. New fields of biological study also have their impact. Bioinformatics, which is a technique of information processing combined with statistics, applied mathematics, biochemistry, and artificial intelligence, makes comparative genomics much easier than it would be otherwise due to the sheer amount of data in any given organism's genome. Rapid advances now being made in the life sciences and in medicine rest on principles derived from an understanding of evolution.

Scientists do have questions, and often disagree, about the patterns and processes of evolution, or "how" evolution occurs. As the fundamental nature of science dictates, scientists' conclusions and explanations about the patterns and processes are tested by experiment, observation, and communication to a professional audience of peers. However, subjecting the "how" of evolution to the scientific process does not call into question "whether" evolution has occurred.

History of Evolutionary Thinking

Most people equate the beginning of evolutionary thinking with Charles Darwin's voyage to the Galápagos Islands in 1831. Yet questions of "when" and "how" living things came to be as we see them were being debated almost 200 years before Darwin made his voyage. Follow the timeline of evolutionary thinking in Figure 1 (p. 4).

Topic:
Evolution

Go to:
www.SciLinks.org

Code: ETK002

The History and Science of Evolution

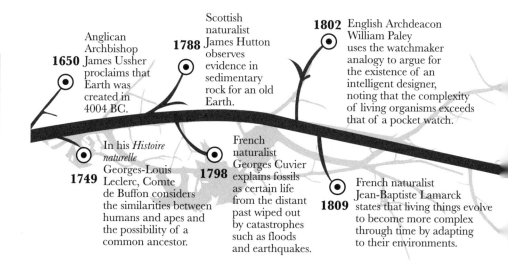

Figure 1. Timeline of Evolutionary Thinking

1650 Anglican Archbishop James Ussher proclaims that Earth was created in 4004 BC.

1788 Scottish naturalist James Hutton observes evidence in sedimentary rock for an old Earth.

1802 English Archdeacon William Paley uses the watchmaker analogy to argue for the existence of an intelligent designer, noting that the complexity of living organisms exceeds that of a pocket watch.

1749 In his *Histoire naturelle* Georges-Louis Leclerc, Comte de Buffon considers the similarities between humans and apes and the possibility of a common ancestor.

1798 French naturalist Georges Cuvier explains fossils as certain life from the distant past wiped out by catastrophes such as floods and earthquakes.

1809 French naturalist Jean-Baptiste Lamarck states that living things evolve to become more complex through time by adapting to their environments.

Around 1650, Bishop James Ussher made meticulous calculations using data from the Bible to set the date of creation of Heaven and Earth as 4004 BC. (Many Young Earth creationists today subscribe to a similar view.) Others, however, looked more closely at the natural world for clues.

One such naturalist was Georges-Louis Leclerc, Comte de Buffon. In the mid-1700s, Leclerc wrote *Histoire naturelle*, a description of everything known in his day about the natural world accompanied by overarching theories about Earth and life on the planet. Contrary to accepted views of his day, Leclerc estimated that the process for Earth's formation—from a comet striking the Sun—took more than 70,000 years. Although a far cry from the age of Earth currently accepted, that number was almost inconceivable to the populace of the day.

Another naturalist, James Hutton, hypothesized that geological processes of the past occurred at rates similar to those he could observe in his own time, making it possible to estimate the time it took, for example, to deposit sandstone of a given thickness. Such analysis made evident that enormous

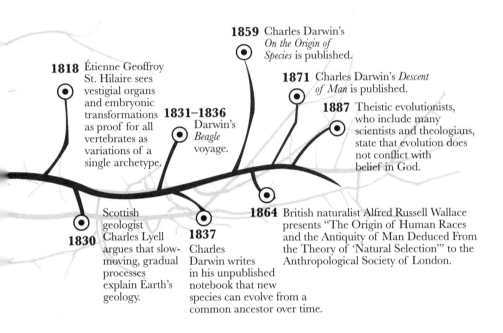

lengths of time were required to account for the thicknesses of exposed rock layers. This concept of an "old" Earth was revolutionary at the time. Today, his idea of uniformitarianism is one of the fundamental principles of Earth science.

Georges Cuvier wasn't the first person to observe that fossils were remnants of past life-forms, but his uncanny ability to reconstruct organisms from fossils led to his understanding that groups of organisms had become extinct. His careful study of fossils showed that many large mammals, such as mammoths and giant sloths, were separate species from similar animals living during his day. Cuvier therefore hypothesized that species became extinct during environmental catastrophe. Geologic history, which shows more constant background extinctions punctuated by mass extinction events, supports his idea.

With evidence of extinct life-forms established, Jean-Baptiste Lamarck thought about "how" life on Earth changed. His hypothesis

The History and Science of Evolution

Figure 1. Timeline of Evolutionary Thinking *continued*

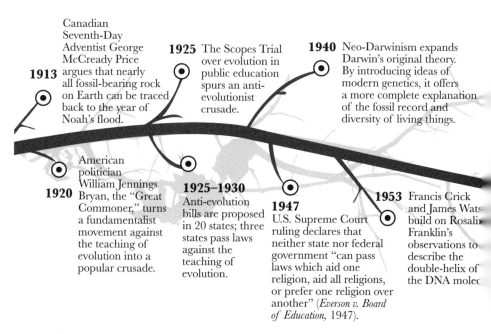

1913 Canadian Seventh-Day Adventist George McCready Price argues that nearly all fossil-bearing rock on Earth can be traced back to the year of Noah's flood.

1920 American politician William Jennings Bryan, the "Great Commoner," turns a fundamentalist movement against the teaching of evolution into a popular crusade.

1925 The Scopes Trial over evolution in public education spurs an anti-evolutionist crusade.

1925–1930 Anti-evolution bills are proposed in 20 states; three states pass laws against the teaching of evolution.

1940 Neo-Darwinism expands Darwin's original theory. By introducing ideas of modern genetics, it offers a more complete explanation of the fossil record and diversity of living things.

1947 U.S. Supreme Court ruling declares that neither state nor federal government "can pass laws which aid one religion, aid all religions, or prefer one religion over another" (*Everson v. Board of Education*, 1947).

1953 Francis Crick and James Watson build on Rosalind Franklin's observations to describe the double-helix of the DNA molecule.

of inheritance of acquired traits has since been rejected, but in his time, it was a coherent and persuasive presentation of the idea that organisms change in response to their environment. In fact, Darwin wrote in 1861:

> Lamarck was the first man whose conclusions on the subject excited much attention. This justly celebrated naturalist first published his views in 1801. . . he first did the eminent service of arousing attention to the probability of all changes in the organic, as well as in the inorganic world, being the result of law, and not of miraculous interposition. (UCMP Berkeley)

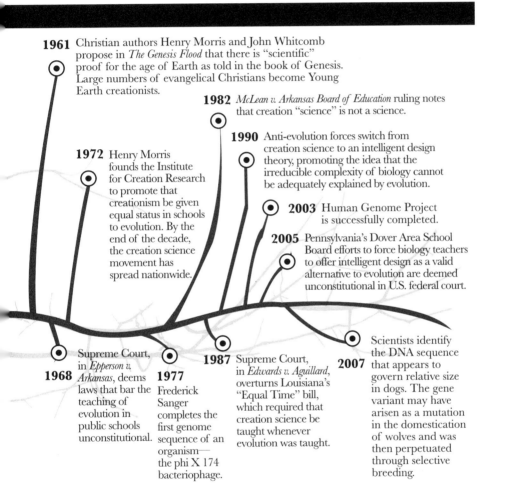

1961 Christian authors Henry Morris and John Whitcomb propose in *The Genesis Flood* that there is "scientific" proof for the age of Earth as told in the book of Genesis. Large numbers of evangelical Christians become Young Earth creationists.

1972 Henry Morris founds the Institute for Creation Research to promote that creationism be given equal status in schools to evolution. By the end of the decade, the creation science movement has spread nationwide.

1982 *McLean v. Arkansas Board of Education* ruling notes that creation "science" is not a science.

1990 Anti-evolution forces switch from creation science to an intelligent design theory, promoting the idea that the irreducible complexity of biology cannot be adequately explained by evolution.

2003 Human Genome Project is successfully completed.

2005 Pennsylvania's Dover Area School Board efforts to force biology teachers to offer intelligent design as a valid alternative to evolution are deemed unconstitutional in U.S. federal court.

1968 Supreme Court, in *Epperson v. Arkansas*, deems laws that bar the teaching of evolution in public schools unconstitutional.

1977 Frederick Sanger completes the first genome sequence of an organism— the phi X 174 bacteriophage.

1987 Supreme Court, in *Edwards v. Aguillard*, overturns Louisiana's "Equal Time" bill, which required that creation science be taught whenever evolution was taught.

2007 Scientists identify the DNA sequence that appears to govern relative size in dogs. The gene variant may have arisen as a mutation in the domestication of wolves and was then perpetuated through selective breeding.

Source: Information in the timeline adapted from
www.pbs.org/wgbh/evolution/religion/revolution/index.html
www.ucmp.berkeley.edu
http://evolution.berkeley.edu/evosite/history
www.wku.edu/~smithch/index1.htm, http://chemistry.nobel.brainparad.com/frederick_sanger.html
www.nih.gov/news/pr/apr2007/nhgri-05.htm

The History and Science of Evolution

Zoologist Étienne Geoffroy St. Hilaire studied morphology with little emphasis on the evolutionary history of forms. His observations, however, identified and defined homologous structures, primary tools of later evolutionary biologists. Not all of Geoffroy's hypotheses stood the test of time. Still, his understandings contributed to others' thoughts about the process of evolution.

Geologist Charles Lyell wanted to find a way to make geology a true science of its own. He disagreed with his mentor William Buckland, who tried to link catastrophism—the idea that geologic landforms and species appeared during sudden upheavals—to Noah's biblical flood. Lyell pursued geologic study that was built on observation, not susceptible to wild speculations, and independent of the supernatural. He gathered evidence of gradual change throughout Europe, building on Hutton's idea of uniformitarianism.

Some of these hypotheses may not have stood the test of time and science, but they all represent early attempts to conceptualize and articulate ideas that Darwin was able to further. He was greatly influenced by the observations and hypotheses of many of these men as he set sail on a winter day in 1831.

Darwin's Thought Process

Topic: History of evolution

Go to: *www.SciLinks.org*

Code: ETK003

Darwin's observations in the Galápagos Islands themselves were a relatively small part of his work on the overall voyage. For more than three years Darwin observed the geology and life of South America as the HMS *Beagle* stopped at ports from Brazil to Chile. During the voyage Darwin was known to have written to Lyell, whose view of landform development enabled Darwin to decipher the history of the Canary Islands. Lyell's ideas influenced Darwin so strongly that he hypothesized a process in which evolution took place from one generation to the next, but too slowly for us to perceive.

Darwin seemed to be more interested in biological changes than geological ones, however, and his five-week survey of life on the Galápagos, as the *Beagle* visited and revisited each island, is well documented. His observations of variations among the individuals in the populations of tortoises, finches, and other inhabitants, and his thoughts about the similarities between these organisms and those on the South American continent fueled his thinking for the rest of the voyage.

Figure 2. Darwin's Survey of the Galápagos Islands

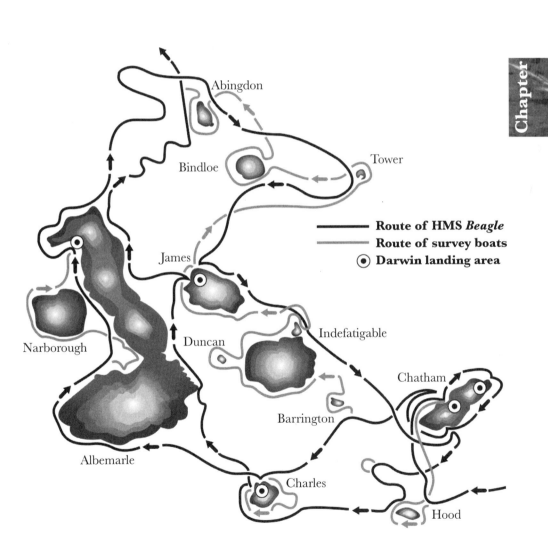

The History and Science of Evolution

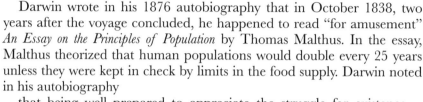

Darwin wrote in his 1876 autobiography that in October 1838, two years after the voyage concluded, he happened to read "for amusement" *An Essay on the Principles of Population* by Thomas Malthus. In the essay, Malthus theorized that human populations would double every 25 years unless they were kept in check by limits in the food supply. Darwin noted in his autobiography

> that being well prepared to appreciate the struggle for existence which everywhere goes on from long-continued observation of the habits of animals and plants, it at once struck me that under these circumstances favourable variations would tend to be preserved, and unfavourable ones to be destroyed. The results of this would be the formation of a new species. Here, then I had at last got a theory by which to work. (Darwin 1959, p. 120)

In the essay, Malthus observed that in nature plants and animals produce far more offspring than can survive and that the same principle applied to humans. Although Malthus ascribed the potential outcome to God's influence, Darwin applied the concept to his observations of the natural world and arrived at a different mechanism for driving evolution—natural selection. Like farmers who practiced artificial selection, nature would "select" the traits most favorable to survival in that particular environment. With an overproduction of offspring, only individuals with certain characteristics or variations would survive to reproduce.

Topic:
Darwin

Go to:
www.SciLinks.org

Code: ETK004

During the time that Darwin pondered his theory before publication, a younger British naturalist, Alfred Wallace, collected biological specimens in Southeast Asia to sell to museums. He sent an essay to Darwin that, surprisingly, came to the same conclusions about how evolution could occur. While Darwin willingly shared credit with Wallace for his theory of evolution by natural selection, review of Darwin's early notes shows much of his thinking was completed while Wallace was a very young man, a fact not lost on Wallace who is said to have given priority and credit to Darwin even after Darwin's death.

The Modern Synthesis

Many textbooks and reference materials end the discussion of evolution with Darwin's work and with descriptions and activities that emphasize natural selection as *the* mechanism of evolution. However, while Darwin convinced scientists of his day that life had and continued to evolve, problems remained with the concept of natural selection. Darwin was bothered by his own inability to understand inheritance, and therefore explain heritable variation, a key requirement for evolution by natural selection. Gregor Mendel, who became known as the father of genetics but who was not yet 10 years old when Darwin set sail on the HMS *Beagle*, would later demonstrate particulate inheritance. But Darwin would never know the potential impact of Mendel's work on his own.

Later scientists did not miss the opportunity to further test the theory of evolution by natural selection in light of this new field of genetics. In the 1930s, Soviet-born scientist Theodosius Dobzhansky studied mutations and made connections to the work of theoretical population geneticists Sewell Wright and Ronald Fisher to determine how the populations of a species differed. Dobzhansky hypothesized that speciation occurred when neutral mutations spreading within a population resulted in that population being unable to interbreed with another population. The work of these scientists, specifically the applying of knowledge from different areas of science to what was known about evolution, became the basis of the modern theory of evolution.

The modern synthetic theory of evolution deviates from Darwin's focus on individuals and speciation. Rather it concentrates on how evolution works at the level of genes, phenotypes, and populations and posits that

- natural selection is just one of several mechanisms at work,
- variation within a population is understood to be due to the presence of multiple alleles of a gene, and
- while speciation is most often due to the gradual accumulation of small genetic changes, evolution can occur very rapidly.

Scientists do not debate whether evolution occurred, but they still seek answers to exactly how the mechanisms work and how quickly it occurs—gradually, in short bursts followed by periods of seemingly little change, or a combination of the two. As we come closer to understanding exactly what combination of factors results in evolution of new species, we will come closer to understanding its gradual or punctuated manifestation.

With James Watson and Francis Crick's conceptualization of the deoxyribonucleic acid (DNA) molecule, most scientists began focusing on genes, nucleotide base pairs, and the impact of mutations on genes. This intense effort led to the first sequencing of an organism's genome—a bacteriophage's 11 genes—in 1977 and to the completion of the Human Genome Project in 2003. The study of genomics continues today. Genomes enable compari-

The History and Science of Evolution

sons among species at the most discrete level. Analyzing these data shows how many base pairs differ between two species and thus how recently they descended from a common ancestor. Analysis also shows just how much genetic variation exists in nature. The 2007 mapping of geneticist Craig Venter's 46 chromosomes has shown variation among humans is five to seven times greater than previously thought.

Most recently, developmental biology has gained new prominence in the study of evolutionary mechanisms. Spurred on by Stephen Jay Gould's research in the late 1970s, the study of evolutionary developmental biology, called *evo-devo* for short, uses embryos to reveal startling similarities across major taxa that are clearly the result of shared ancestry. The central tenant of evo-devo is that a set of "tool kit genes," common to most animals, directs the development of corresponding parts such as eyes or appendages in each species. Whether or not a trait exists depends on whether the gene is expressed, with expression often regulated by noncoding regions of DNA (segments that do not code for a specific protein) that act as "switches," turning on or off a gene's expression.

Paired box gene 6, or Pax6, is one such gene that regulates transcription processes that control the development of eyes and other sensory organs. Pax6 has been identified in such diverse organisms as mice, fruit flies,

Topic:
Genome research

Go to:
www.SciLinks.org

Code: ETK005

Evidence for Evolution

As with all scientific theories, the evidence for evolution is overwhelmingly pervasive and connected to many areas of science. The evidence for evolution is constantly accumulating in the scientific literature as new work proceeds and new discoveries are published. The following websites provide additional background information on evolution, which should be helpful as you develop your lesson plans.

Fossil record
Although incomplete, the fossil record clearly shows how life has changed over the past four billion years. *Tiktaalik roseae*, a fossil fish discovered in the arctic in 2006, provides an excellent example of the predictive power of evolution. *http://tiktaalik.uchicago.edu/index.html*

Comparative anatomy
Use the examples of vertebrate arms likely found in your textbook materials as homologous structures. Emphasize that although shared patterns are best explained by inferring their inheritance from a common ancestor that also had this pattern, homologous structures in and of themselves are not evidence

of common ancestry. The homologous nature of structures is a hypothesis that is evaluated and tested given the evidence of comparative anatomy, genetics, development, and behavior. Common ancestry is inferred based on these sources of information and reinforced by the patterns of similarity and dissimilarity of anatomical structures. Review this background on homology to enhance your discussion.
www.ncseweb.org/icons/icon3homology.html

Biogeography
The study of the distribution of flora and fauna around the world provides clues to the process of evolution. Biogeography links the migration and isolation of populations to geologic events and ensuing geographic barriers. Marsupials typify these links. Consider their dominance of Australia with only a few species in South and North America. Relate that to the fact that Australia and South America were once part of Gondwanaland. After that landmass's breakup, continental drift isolated Australia. Marsupials flourished. What happened on the South American continent? You might research other examples such as desert cacti, pineapples, or horses. This blog entry describes what biogeography can tell you about the relatedness of American and European web-toed salamanders.
http://blog.jcmnaturalhistory.com/?p=191

Molecular genetics
Investigating genes at the molecular level is currently an area of active scientific research. Comparing the sequences of similar proteins isolated from different species generates daily contributions to the mass of evidence for evolution. For example, one protein called *cytochrome c* has been examined in more than 80 species. These cytochrome c amino acid sequences can be used to quantify differences between species. Evolutionary trees constructed from comparisons of proteins are much like those based on comparative anatomy.
www.talkorigins.org/faqs/molgen

Developmental biology
Combining the study of evolution and the development of living things—from embryo through adulthood—through the lens of molecular biology is shedding new light on all aspects of speciation. Look to the *New York Times* for an overview of the potential impact of evo-devo.
www.nytimes.com/2007/06/26/science/26devo.html

humans, and squid. In one study, mouse Pax6 triggered eye formation in fruit flies, evidence that the common developmental mechanism at work is the result of shared ancestry rather than convergent evolution among unrelated species. The fact that a relatively small number of tool kit genes are found in most animal species strongly suggests descent from a common ancestor. Similar tool kit genes are found in plant species as well.

The History and Science of Evolution

The Genetics of Evolution

What causes genetic variation? In eukaryotes, mutations and genetic recombination primarily cause genetic variation. Mutations are random, resulting from mistakes during replication or induced by radiation or chemicals in the environment. Mutations happen. Most mutations are neutral, having little or no effect. Natural selection operates on beneficial and unfavorable mutations, either maintaining beneficial mutations or removing unfavorable mutations. Most mutations that persist cause small changes, although they can add up to large changes over time. Sometimes mutations cause more noticeable changes. A mutation in the gene that helps make hemoglobin changes the shape of the hemoglobin molecule, allowing it to clump together. This results in a sickle-shaped red blood cell. Individuals with one copy of the mutated gene, which causes half of their red blood cells to be sickle shaped (sickle-cell trait), have greater resistance to malaria than those individuals without the mutant gene, who have all normal-shaped red blood cells. In a malaria-prone environment, individuals with the sickle-cell trait are more likely to pass along their genes than those without, so the gene stays in the population.

Topic:
Species/
speciation

Go to:
www.SciLinks.org

Code: ETK006

Mutations can also result in the rearrangement of chromosomes. Humans have 23 pairs of chromosomes, one less pair than chimpanzees, gorillas, orangutans, and other great apes. For about 20 years, researchers hypothesized that human chromosome 2 resulted from the fusion of two mid-sized chromosomes found in these apes. Now, studies show DNA sequences that confirm the location of the inactivated centromere, or area where the arms of the chromosome are attached.

Genetic recombination occurs in sexually reproducing populations as the meiotic process enables different genes to come together. The larger the population and the more random the mating, the greater the shuffling of gene pairs that will result. Different gene pairs will result in differing phenotypes, but recombination does not result in new genes being added to the gene pool as mutations do.

Each individual in a population has a unique set of alleles, or forms of specific genes, that determine its physical features, behaviors, and other characteristics. Natural selection acts on each individual as a whole, not just on the genes that determine any given trait. But if the individual is successful at surviving long enough to reproduce, then its alleles are passed on, including any that might impede survival if singled out. If the individual dies before reproducing, its alleles are lost to the gene pool. Because these alleles are carried by other individuals in the population, change does not depend on whether any particular individual does or does not pass on its genes. For this reason, evolutionary biologists focus on the population rather than the individual in trying to understand the process of evolution.

Laws of probability predict the frequency with which alleles are present in a population. Yet, in a small population the frequency of a given allele may be much higher or lower than expected. This genetic drift occurs when individuals carrying a particular allele do or do not reproduce, not because the allele is useful or not useful in the environment, but just because of chance. Such chance changes in gene frequencies are more likely to be important in small populations. Genetic drift can happen when a small group of individuals colonize a new habitat. The founding individuals may carry alleles that are less common than others found in the original population, yet they will be very common in the new population.

Migration also affects the frequencies of alleles in a given gene pool. When a group migrates, its alleles are lost to the original population but gained by the new population. These changes can be significant in small groups where an influx of "new" genes can change the frequency of those genes and thus the potential for natural selection or drift. Isolation—geographic, behavioral, or temporal—subjects the gene pool to stagnation except for mutations.

All of these mechanisms contribute to speciation in sexually reproducing organisms. If a population's gene pool changes to the point where individuals cannot viably reproduce with individuals of the same species in another population, biological speciation occurs. These mechanisms do not affect asexually reproducing populations in the same way. Therefore, the concept of "species" is not necessarily a hard-and-fast one. Depending on a researcher's focus, other factors influence any given delineation of a species, such as mate recognition, niche, similarities in DNA, phenotype, and morphology.

Chapter 2:
Active Evolution Instruction

Craig Nelson, biology professor at Indiana University and researcher in evolution and evolutionary ecology, asked the following question in his book *The Creation Controversy & The Science Classroom* (Skehan and Nelson 2000, p. 19): "How can we produce a scientifically literate society, especially in areas that are publicly controversial?" He came to the following conclusions:

- "Active learning is even more important for controversial topics than for the rest of science."
- "We too often teach science as a set of conclusions … instead of as a set of processes for thinking critically about alternatives."
- "We have good ways to judge the levels of strength of support for scientific theories and other criteria for comparing them. Helping students understand these ways of judging is integral to teaching science as critical thinking."
- "Public controversies usually rest on disagreements about consequences … If students are to understand why topics such as evolution are controversial, we must help them understand the different views of consequences."

In his book, Nelson also describes effective strategies for teaching evolution and other controversial topics by outlining several problems and giving strategies for how to overcome them. Using evolution instruction as his example, Nelson applies strategies that emphasize critical thinking and that can be adapted to your classroom situation.

Research shows…
- The public's view of the goals of science education may differ from your personal teaching goals. The public rated the following goals of science education as "very important":

Learn how to draw conclusions from evidence	80%
Learn how to think critically	78%
Learn how science is conducted	63%

  (Coalition of Scientific Societies 2007, p. 6)

- Biology teachers whose college-level coursework included at least one class in evolutionary biology devote substantially more class time to evolution than teachers with fewer credit hours in evolutionary biology. The best-prepared teachers devote 60% more time to evolution than the least prepared (Berkman, Pacheco, and Plutzer 2008, Table S6).

Active Evolution Instruction

To help elevate your level of active instruction, this chapter provides a sampling of the kinds of hands-on activities that promote understanding of evolutionary processes. These particular activities are structured according to the BSCS 5E Instruction Model—engage, explore, explain, elaborate, and evaluate. The 5E Model is a teaching sequence that can be used for entire programs, specific units, or individual lessons (Bybee 1997). This chapter also contains print and online resources for new ideas and practical applications that will help you emphasize science process and how evolutionary changes affect us daily.

Showing Evolutionary Relationships With Cladograms

Engage

As you introduce the concept of comparative anatomy reflecting common ancestry, your program or reference materials might suggest a common classification task whereby students group inanimate objects, such as classroom furniture, lab equipment, or backpack contents, according to visible characteristics. However, this kind of task can lead to misconceptions because unlike living things, there is no genetic relationship among the objects used. Different lab groups will develop different classification systems, but as long as the logic is sound, no one system will be any more "correct" than the next.

Instead of this common classification task, show students pictures of organisms that have several common characteristics such as streamlined swimming marine animals—jellyfish, tuna, swordfish, killer whale, dolphin, great white shark, and stingray. Have students group the animals and then ask volunteers to share their grouping criteria. Ask students how the characteristics of the animals are determined (gene interaction) and what processes would have to occur for their characteristics to change (mutation, adaptation). Help students see how a desk, for example, may have different physical characteristics in different circumstances that would lead to alternate classifications. Great white sharks, though, are great white sharks no matter the circumstances and have a definite set of characteristics.

Explore

Have students consider the "relatedness" of a subset of these marine animals. Ask students to consider which animals are more closely or more distantly related. Focus students on the jellyfish, tuna, killer whale, and great white shark. To differentiate among the animals, have students determine whether the animals' skeletons are cartilaginous or bony and whether breathing is accomplished through gills or lungs. Develop a class chart with characteristics listed across the top and the names of the animals down the side (see Table 2). Have students research the characteristics and use symbols such as + and 0 to denote if a character is present in a given animal at any time during its development.

Table 2. Characteristics of Selected Organisms

	Vertebrae	Bony skeleton	Lungs
Jellyfish	0	0	0
Tuna	+	+	0
Killer whale	+	+	+
Great white shark	+	0	0

Explain

Tell students that the modern process of classification is based on phylogenies. A phylogeny is similar to a family tree that shows how the various types of organisms in a group are related. Comparative anatomy studies and homologous structures will show characteristics that are the same for all types of organisms in the phylogeny. These characters are the ancestral ones that stem from the common ancestor. Shared derived characters, or ones that differ from the ancestral characters, show the evolutionary pathway. By drawing a diagram of branching lines that connect the groups that share the derived character, the degree of relationship can be inferred. These diagrams look like trees and are called phylogenetic trees or cladograms. (The word *cladogram* originates from the Greek word *klados*, meaning branch or twig.) Reproduce the cladogram shown in Figure 3 (p. 20) for students.

Use the class chart to explain how the cladogram is developed. First, note that all organisms except the jellyfish have vertebrae. Therefore, you

Active Evolution Instruction

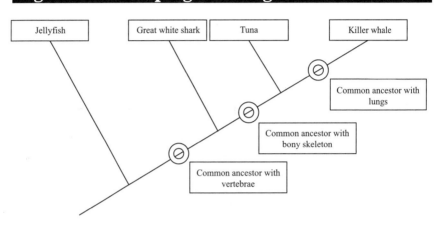

Figure 3. Developing a Cladogram

would designate the jellyfish as the "out-group," and it branches off the cladogram first. The remaining three animals (great white shark, tuna, and killer whale) all descended from an ancestor with vertebrae.

Consider the next characters, the presence of a bony skeleton and lungs. Have students determine which characteristic is shared by all remaining animals except one. The great white shark has a cartilaginous skeleton, whereas the tuna and killer whale have bony skeletons. So the shark is the next animal to branch off the cladogram. Because both tuna and killer whales have bony skeletons, it is most likely that they share a common ancestor with a bony skeleton. Finally, point out that the killer whale has a characteristic that is never present in the other organisms—lungs. Therefore, the killer whale is positioned at the end of this cladogram. In this particular example, lungs cannot be considered a shared derived character because the killer whale is the only animal in the chart displaying the character. Lungs are, however, a derived trait that evolved in the lineage leading to killer whales. Lungs would be a *shared* derived trait if the chart included dolphins because lungs have been inherited from the common ancestor of killer whales and dolphins.

Help students understand how drawing branches from the ancestral line is important. If the organisms were located on the ancestral line it would give the impression that sharks evolved into tuna, which then evolved into killer whales. Although students sometimes misunderstand Darwin's concept of descent with modification, in this model students can see that the "modification" is a derived character that branches from the ancestral line, rather than the "modification" of one group into another. Sharks, tuna, and killer whales descended from a common ancestor, but did not evolve directly from one into the other. Also be sure that students do not think that because the killer whale is at the end of the straight line, it is the common ancestor from which the other types of animals evolved.

The cladogram also shows the relative passage of geologic time. Point out to students that when reading a cladogram, the longer the line, the more distant in the past the event occurred. You might draw the cladogram reversed from right to left, keeping all relative line lengths the same for comparison. Because we read left to right, students may have trouble seeing that the two versions show the same relationship with regard to time—that the common ancestor of tuna and killer whales occurred more recently in geologic time than the common ancestor of tuna, killer whales, and great white sharks.

Cladograms that show several branches can sometimes be confusing to students. Draw the cladogram shown in Figure 4 to spark discussion.

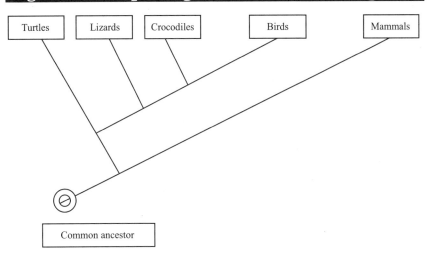

Figure 4. Interpreting Branches of a Cladogram

Topic:
Phylogenetic trees

Go to:
www.SciLinks.org

Code: ETK007

Ask students whether birds are more closely related to mammals or to turtles. Help students see that the proximity of the labels is not an indication of relatedness. Instead, students should look at the point of branching.

Active Evolution Instruction

Birds and turtles share a more recent common ancestor than birds and mammals. To clarify, you could draw the same cladogram, but draw the branching from the turtle lineage to the left, which would reverse the order of the labeling but not change the comparative evolutionary relationships of the organisms. On the cladogram, close relationships are shown by a recent fork from the supporting branch. The closer the fork in the branch between two organisms, the closer their relationship. Help students see that birds are most closely related to crocodiles and more closely related to lizards than they are to either turtles or to mammals.

Elaborate

To visualize how characteristics can show evolutionary relatedness, use a set of five nested (large-to-small) self-sealing plastic bags, either for a demonstration or provide one set per lab group. (Six sizes for nesting will be needed by the end of this Elaborate stage.) Make a set of "animal" cards for each set of bags. Label the cards *lamprey, perch, turtle, kangaroo,* and *mouse*. If students are unfamiliar with these animals, have them find pictures and research general characteristics of each.

Also prepare a set of "character" cards using the characteristics in Table 3 for each set of bags (one card per column). If possible, make each character card a different color or add a colored marker symbol for easy reference later. Reproduce Table 3 on a blackboard, white board, or overhead—any means by which students can add to the chart later. You should point out to students that though the characteristics in Table 3 are grouped together, they are inherited independently of one another.

Table 3. Characteristics for Character Cards

Character ○	Character ◊	Character □	Character Δ	Character +
Dorsal nerve cord Notochord Chambered heart	Jaws Paired appendages Vertebral column	Amniotic egg	Three ear ossicles Hair Mammary glands	Placenta

Spread out the five plastic bags from largest to smallest. Near the top, label the bags A through E from largest to smallest. Attach the character card with the characteristics common to the greatest number of organisms [O] to the largest bag (Bag A). Place the organisms with those characteristics inside that bag. (All go into Bag A.) Then attach the character card with characteristics common to the next greatest number of organisms [◊] to the second largest bag (Bag B). From Bag A, take as many organisms that have those characteristics and move them to Bag B. (All except the lamprey go into Bag B.) Repeat until all the character cards and animals have been distributed among the bags (Bag C, [□], lizard; Bag D, [Δ], kangaroo; Bag E, [+], mouse). Now "nest" the bags so they are inside one another and the character and animal cards are visible.

Lead students to understand that the characteristics on any given bag are not only characteristics of the animal in that bag, but also of all the other animals in the bags nested within. Any derived character is also a characteristic of those animals nested within that bag, but not of the ones outside it.

Now give students a "frog" card. Have students research the unique character set of this organism (lungs, four limbs) and add that data to Table 3. Ask students where a new bag with these characters would nest within their set (between Bags B and C). Give students an appropriate size bag to insert.

Have students convert their visual Venn diagram of nested bags to a chart similar to the one you made during the Explore stage. Have students list the character sets across the top and the animals down the side. Students should complete the chart using Xs to show whether that organism has that trait or not. Have students compare the total number of Xs for any given animal to its bag's position in the nested arrangement. Use the chart to revisit the concept that any derived character is also a characteristic of those animals nested *within* that bag, but not of the ones *outside* it.

Next, ask students to draw a cladogram using the chart and their nested bags as a guide. Once students have constructed their cladograms, they can map the characters onto the cladograms as in the previous examples. Troubleshoot as students work independently or in small groups ensuring that time is illustrated correctly and the number of shared characteristics increases appropriately. Show the example in Figure 5 (p. 24) to students who are struggling with the material.

Evaluate

Tell students that the Hawaiian Islands are home to at least 800 species of *Drosophila*, or fruit flies. Show students pictures or describe the fragility of fruit flies and how a fruit fly's range would cover a relatively small

Active Evolution Instruction

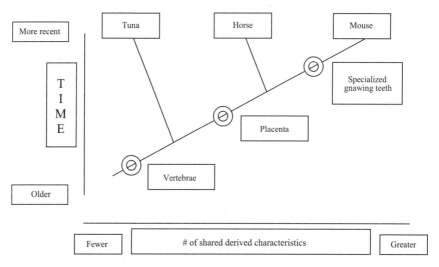

Figure 5. Sample Cladogram Emphasizing Time

geographic area. Scientists think that one pregnant fruit fly blew ashore on one of the islands several million years ago. This one fruit fly is the common ancestor of the hundreds of species on the islands. Give students the cladogram in Figure 6, developed from the ancestral and shared derived characters of five species of fruit flies. Each species lives on a different island in the chain.

Display a map of the Hawaiian Islands and have volunteers research and share basic facts about the formation of the Hawaiian Islands over a mid-plate hotspot and their ages. Add island names and approximate formation times (Niihau and Kauai formed about five million years ago; Hawaii formed about one half million years ago).

Have students write two to three paragraphs that explain how the islands' formation supports the hypothesis of the evolutionary relationships shown by the cladogram. Writing should reflect an understanding of the relative length of evolutionary history of each species, potential islands that each species lives on, the relative time over which the speciation occurred, and how the islands' origins might have contributed to speciation.

To assess students' responses, use the SAT Reasoning Test Essay Scoring Guide (http://professionals.collegeboard.com/testing/sat-reasoning/scores/essay/guide), the Northwest Regional Educational Laboratory's 6+1 Trait Assess-

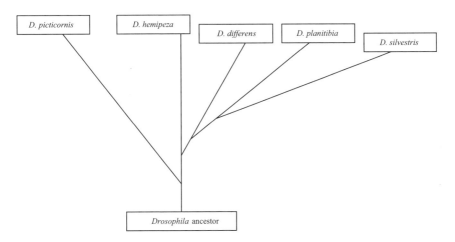

Figure 6. Hawaiian Fruit Flies

ment Scoring Guide (*www.nwrel.org/assessment/pdfRubrics/6plus1traits.PDF*), or your own state or local writing rubric. Writing activities help students assimilate ideas and provide practice for the writing required on the SAT and state-mandated exams.

Natural Selection and Antibiotic Resistance

Engage

Engage students in a discussion about hand sanitizers and other antibacterial or antimicrobial soaps. Elicit from volunteers how often they use such products and why they use them over more conventional soaps. Draw on students' prior knowledge to develop an operational definition of antibiotic. Continue the discussion, probing for use of antibiotics and antibacterial products to combat illnesses. Ask students to list illnesses that antibiotics effectively fight. Remind them that viral infections such as colds, flu, and most sore throats should not be treated by antibiotics because viruses are not affected by antibiotics. Point out to students that prescriptions for antibiotics specify that all of the prescription must be taken over a specific timeframe. Ask for hypotheses as to why this is the case.

Active Evolution Instruction

Explore

Set up your favorite predator/prey simulation using multicolored paper dots, toothpicks, jelly beans, or tokens on a colored "environment." Have students demonstrate how a population can shift due to predation. "Prey" are captured (picked up with fingers, forceps, or sticky tape; stabbed with a toothpick; etc.) during a 30-second turn and are not returned to the environment. Students observe the survivors and add "offspring" to the surviving population. For every pair of similar-colored objects not captured, two offspring are born (or added back to the environment). In the event that an odd number of objects survive, students should calculate offspring based on the highest number of pairs that can be formed. The objects that are easy to find will have few survivors and may become extinct due to their low reproduction rate. Those that are difficult to find will survive, reproduce, and increase in total population. After even a few turns, students witness a decline in diversity as the population shifts toward objects that are most similar to the color of their environment.

Explain

Connect students' exploration to their discussion of antibacterial soaps and antibiotics. Guide them to understand that the overuse and misuse of these products enables natural selection to act on bacterial populations resulting in shifts to populations of bacteria resistant to a given antibiotic. Review their hypotheses from the earlier discussion, noting that research shows that a shortened course of antibiotics will kill only the most vulnerable bacteria, while allowing relatively resistant bacteria to survive.

Students may wonder how bacteria evolve so quickly. Describe that at the most basic level, the process of evolving resistance happens quickly for two main reasons:

- Bacteria have large population sizes. Any population is relatively likely to include an individual that happens to carry a gene for resistance.
- Bacteria have short generation times. A population in a patient can evolve from a susceptible state to a resistant one quickly.

Topic:
Antibiotic resistance

Go to:
www.SciLinks.org

Code: ETK 008

Another unusual factor comes into play as well. In a process called *horizontal gene transfer*, bacteria can pass gene copies to one another directly, even to an entirely different species of bacteria.

Many resources, such as the following, describe the process of and health concerns about antibiotic resistance:

Centers for Disease Control and Prevention *(www.cdc.gov/ncidod/eid/vol7no3_supp/levy.htm):* Triclosan is the active ingredient in many antibacterial soaps and it inhibits the production of fatty acids. Research into the particular cellular site on which triclosan acts shows that resistant mutants of *Escherichia coli* (*E. coli*) appeared with low-, medium-, and high-level resistance. They all had a mutation on one gene that enabled the resistant strains to produce essential fatty acids even in the presence of triclosan. In comparison with the wild-type *E. coli*, mutants required up to 100 times more triclosan to show even minimal inhibition of fatty acid biosynthesis.

Although triclosan's high concentration in soaps (2,500 µg/ml) seems to be high enough to kill even resistant bacteria, tests show this is not the case. A 90% death rate in wild-type *E. coli* requires exposure to 150 µg/ml of triclosan in soap for two hours at 37°C. Mutants required two to four times that. Ironically, triclosan was more effective by itself as soaps seemed to decrease triclosan's effectiveness. Mutant strains survived in triclosan soaps diluted with as little as three parts water. Most important, the time, temperature, and amount needed to kill the bacteria greatly exceeded the average five-second hand washing performed by most people.

World Health Organization *(www.who.int/mediacentre/factsheets/fs194/en/*): When infections become resistant to first-line antimicrobials, treatment has to be switched to second- or third-line drugs, which are nearly always more expensive and sometimes more toxic. For example, drugs needed to treat multidrug-resistant forms of tuberculosis are more than 100 times more expensive than the first-line drugs used to treat nonresistant forms. In many countries, the high cost of such replacement drugs means that some diseases can no longer be treated. Most alarming are diseases where resistance is developing for virtually all currently available drugs. Even if the pharmaceutical industry were to step up efforts to develop new replacement drugs immediately, current trends suggest that some diseases will have no effective therapies within the next 10 years.

Mayo Clinic (*www.mayoclinic.com/health/antibiotics/FL00075*): To halt the spread of antibiotic resistance: understand when antibiotics should be used, take antibiotics exactly as prescribed, never take an antibiotic without a prescription, don't pressure a doctor for antibiotics when you have a viral infection, and protect yourself from infection in the first place.

Active Evolution Instruction

Elaborate

For years it was commonplace to have students prepare agar plates, transfer bacteria from the human body and surfaces of objects in the classroom, and add antibiotic discs to show the effect of antibiotics on bacteria growth. Students would then observe any resistant colonies forming in the clear zone around the desks. Now, however, that practice is discouraged. The December 2007 issue of *NSTA Reports* advises against such an activity, because whenever students create bacterial cultures in schools, it is almost inevitable that rich colonies of staphylococcus (along with streptococcus and other pathogens) will be produced (Texley and Kwan 2007). Placed in an optimal medium for 24 hours, a single cell might produce a colony of 10^9 cells/ml by the next day. Instead of having students work with bacteria and antibiotics, show them images that you can find online or in text materials.

Promote critical thinking by giving students the generation times of common pathogens and asking them to hypothesize how the difference in times might affect population shifts (Table 4). Or share with students data on observation of antibiotic resistance (Table 5). Have students research common ailments each antibiotic was used for and hypothesize why resistance was relatively quick or slow to evolve.

As an alternative, students can work with microbes that can live under extreme hypersaline conditions (Schneegurt, Wedel, and Pokorski 2004). The high salt content of the media eliminates the need for sterilization and aseptic techniques, and rules out the possibility of culturing bacterial pathogens. Staphylococcus

Medium for Halotolerant Bacteria

Mix and boil the following ingredients until the agar or gelatin dissolves:
- 50 g (5 tsp) iodine-free table salt (NaCl)
- 1.25 g (1/8 tsp) salt substitute
- 0.5 g (1/8 tsp) all-purpose plant food
- 5 g (3/4 tsp) unsulphured molasses
- 3.75 g (1 1/2 tsp) agar-agar or 33.6 g (12 tsp) gelatin
- 180 ml (3/4 cup) tap water

After cooling the mixture to about 45°C (baby-bottle temperature), add the following:
- 50 ml (10 tsp) of milk solution—2.5 g (1 1/2 tsp) instant nonfat dry milk in 50 ml (10 tsp) tap water
- 20 ml (4 tsp) vitamin solution—dissolve 1 multivitamin tablet in 50 ml (10 tsp) tap water with low heat (<45°C) and decant or filter
- 1.5 ml (5 drops) antifungal preparation (1% clotrimazole solution)

is relatively halotolerant, but it does not grow above 15% (w/v) salinity. High-salt concentrations of 25% or more, and the addition of magnesium (as magnesium sulfate in Epsom salts to 1%), will often enrich for haloarchaea—brightly colored red or pink organisms that depend on high-salt concentrations for growth. Therefore, salt concentrations of 20% to 25% (w/v) should be used in this activity. "Salty Microbiology" in the September 2004 issue of *The Science Teacher* provides a good source of background information on where to find halotolerant species, as well as the medium recipe of approximately 250 ml (1 cup) provide in the sidebar.

Table 4. Pathogenic Bacteria Generation Times

Pathogen	Condition	Generation time
Escherichia coli O157:H7	diarrhea	20 minutes
Staphylococcus aureus	infections	30 minutes
Mycobacterium tuberculosis	tuberculosis	870 minutes
Treponema pallidum	syphilis	1980 minutes

Table 5. Resistance to Antibiotics

Antibiotic	Year first used	Year resistance observed
Penicillin	1943	1946
Streptomycin	1943	1959
Tetracycline	1948	1953
Erythromycin	1952	1988
Methicillin	1960	1961
Ampicillin	1961	1973

Make an analogy between how the extreme environment selects for individuals that can tolerate a higher salt concentration than the rest of the population and those that are no longer affected by antibiotics (Schneegurt, Wedel, and Pokorski 2004).

Active Evolution Instruction

You might also have students conduct internet research to find out about the activities and education efforts of various government and health agencies. As an example of the varied resources available, see the video program developed by the National Institutes of Health that gives a practical application to evolutionary processes in bacteria at *http://science.education.nih.gov/supplements/nih1/diseases/default.htm*.

Evaluate

To demonstrate their understanding of the causes and concerns of antibiotic resistance, have students develop a public service campaign to educate consumers about overuse and misuse of antibiotics and antibacterial soaps. They might develop press releases, posters, cartoons, computer presentations, or videos. Their campaigns should be informative but might also include humor or elements of art. Have students choose an audience such as elementary students, high school graduates, or the elderly as the target of their messages about evolution in bacteria and health concerns about antibacterial resistance. As a reminder, you can assess student campaigns with the tools listed earlier in the cladogram activity (p. 24).

Using Ring Species to Demonstrate Human Evolution

Engage

Topic:
Human evolution

Go to:
www.SciLinks.org

Code: ETK009

Elicit from students a definition of species and have them relate how speciation can occur through geographic isolation. Prompt them to discuss populations that become separated by a physical barrier or a range so big that populations become isolated in certain portions of it. Remind students that groups of organisms become separate species as characteristics determined by and mutations within their genetic makeup promote survival in their environment.

Explore

Introduce the concept of a ring species, or one that occurs when a single species becomes geographically distributed in a doughnutlike pattern over a large area. This is a phenomenon in which variety of a species gradually change around a physical barrier. The species form a ring such that the two ends of the loop are different enough to be reproductively isolated even though elsewhere in the loop adjacent populations do interbreed. Two well-studied examples are salamanders in the *Ensatina eschscholtzii* group, distributed in mountains along the west coast of North America, and greenish warblers *(Phylloscopus trochiloides)*, small, insect-eating songbirds in the forests of Central and Northern Asia and Eastern Europe.

Demonstrate a ring species distribution by having the class form a ring around a row of desks. The row of desks represents a mountain range for the salamanders and a desert for the warblers. Give each student one of a series of paint chips that gradually vary in hue from one color to another distinct color. Use color families that differ from human skin tones, such as greens and blues. Make an analogy between the gradual change in the colors of the paint chips and the gradual changes in the populations that led to two different species. Scientists hypothesize that the species (color) near the center of the range (series of colors) is the founder or ancestral species to both "ends."

Explain

Show students Figure 7 (p. 32)—the cladogram showing the relationship among humans and different species of apes. Point out to students how the cladogram shows that these separate species did not evolve from one into the other, but from separation of populations of a common ancestor. Adapt the ring species demonstration to show the relationship of humans and chimpanzees and to emphasize that humans did not "descend" from apes but rather from a common ancestor.

Have the class form a ring again, with the same color paint chips used in the previous activity. Describe for students how this ring is not formed in "space" (around a mountain range or desert) but in "time" (over millions of years). Then have students move from a loop to a "V" shape. Students should compare this new shape to the cladogram in Figure 7 and draw the conclusion that the apex of the "V" is the closest common ancestor and the distance along each branch of the student "V" represents time.

Help students understand that gradual variations over time along both halves of the loop constitute evolution along the routes to human and chimpanzees. In this case, the intermediates have gone extinct and the ring is discontinuous in the present. However, as the intermediate forms of salamanders are present in the "space" example, over time, intermediate forms of humans and chimpanzees existed.

Active Evolution Instruction

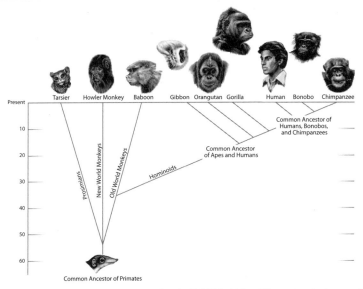

Figure 7. Primate Evolutionary Tree

Source: Biological Sciences Curriculum Study (BSCS). 2005. *The nature of science and the study of biological evolution.* Colorado Springs, CO: BSCS. © 2005 by BSCS. All rights reserved. Reprinted with permission.

Elaborate

Have students compare phenotypes of humans and chimpanzees with baboons, with which they share a much earlier common ancestor, to discern both common and differing structures and behaviors. This task will give students a feel for how many similarities exist between humans and chimpanzees.

Students should then apply their knowledge of DNA to understand the molecular basis for those similarities. Tell students that Svante Pääbo, the director of the Max Planck Institute for Evolutionary Anthropology in Germany, was the first to extract DNA from fossil humans and compare it with living humans and chimpanzees. From this data he estimates that humans and Neanderthals had a common ancestor one half million years ago. Pääbo and his colleague, Henrik Kaessmann, then traced the ancestry of humans and chimps to find out when they shared a common ancestor by using a segment of human DNA about 10,000 nucleotides long from the X chromosome. By counting, they discovered that humans and chimps aver-

Topic:
Species/
speciation

Go to:
www.SciLinks.org
Code: ETK006

aged only about 100 differences out of 10,000 nucleotides on that length of DNA—a difference of about 1%. The scientists concluded from this data that the common ancestor of humans and chimps existed in Africa about five to six million years ago. Use the activities on pages 34 through 39, from *Virus and the Whale: Exploring Evolution in Creatures Small and Large* (Diamond 2006), to show what DNA comparisons can tell us about evolutionary relationships.

Evaluate

To evaluate student understanding of DNA, ask them to write a short news story about humans and chimpanzees as an "assignment" for their local newspaper. In their articles, students should tell their readers about how new DNA studies of humans and chimpanzees suggest they are close relatives. As a reminder, you can assess student writing using the tools listed earlier in the cladogram activity (p. 24).

I cannot accept the fact that I am descended from an ape; therefore, I do not support the theory of evolution.

Evidence does not support the assertion that humans evolved from an ape. Evidence does, however, indicate that modern apes and humans are closely related and are descendents of a common ancestor.

Active Evolution Instruction

It's Molecular Time

DNA is no ordinary stuff. DNA carries the recipe for assembling proteins into living organisms. This long molecule also acts like a clock, helping scientists estimate how long two species have been separated from a common ancestor. When DNA makes a copy of itself, it isn't always perfect. Mistakes can happen. One nucleotide (Adenine, Thymine, Cytosine, or Guanine) might be missing, duplicated, or two nucleotides might switch positions. Scientists like Svante Pääbo have discovered that these mistakes (mutations) accumulate at a regular rate over millions of years, like the steady tick of a clock. Scientists can use this knowledge to date different copies or generations of DNA. This is what scientists call the "molecular clock."

See if you can tell time using the molecular clock method. First, compare copies of tiny segments of DNA for differences (mutations) in the nucleotides. Then use the mutation rate to date the copies.

Work With a Partner

Each team will need:
- Generations of Copies sheet
- scissors

1. Down to DNA

a. Below is a short segment of DNA that is nine nucleotides long. The first line of nucleotides is from a living organism. The second line is from a close ancestor. Underline the nucleotide in the ancestor DNA line that is different from the living DNA. This difference in the DNA code is a mutation site.

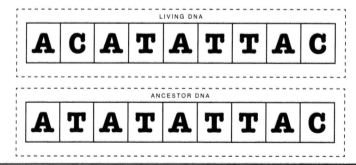

Pages 34 through 39 from Diamond, J., ed. 2006. *Virus and the whale: Exploring evolution in creatures small and large.* Arlington, VA: NSTA Press.

The letters represent nucleotides:

A = Adenine

T = Thymine

C = Cytosine

G = Guanine

2. Sorting Out Sequences

a. Cut out all the sequences of DNA from the Generations of Copies sheet.

b. You have a nine-nucleotide sequence from a section of DNA similar to what you might find from a living human. Look for the sequence with the label living DNA.

c. Now find the closest ancestor. It is the sequence that differs by only one nucleotide. The remaining sequences you have cut out are each older ancestors. Each older ancestor has more mutations or differences.

d. Sort the sequences in order from the living human (the present) to the oldest ancestor (longer and longer ago).

3. Consider This

Now make the DNA tell time. Assume that the rate of mutation in this DNA segment is one difference for every 10,000 years.

a. Based on the rate of mutation, how many years different are the oldest and the newest DNA segments?

b. How long ago did the oldest ancestor live?

Active Evolution Instruction

Generations of Copies
Below is a sequence of nucleotides from DNA that is similar to what you might find from a living human (Living DNA). The rest are sequences from five different ancestors. Cut out the sequences and sort the sequences starting with the living human to the most distant ancestor.

Mutations Up Close

Scientists like Pääbo line up the DNA sequences of different species. They compare the nucleotides A, T, C, G, letter for letter, and count the differences. The differences are the number of mutations in the DNA code that have accumulated over time. The more differences that accumulate between species, the longer the species have been evolving separately.

The genetic code of humans and chimps is billions of letters long. Working out a method for comparing DNA sequences between the two species is one of the problems that genetic scientists must solve. Today this is your challenge too. Only about 1% of the DNA in the chimp and human genes is different. Can you pinpoint the differences?

Work With a Partner

Each team will need:
- Chimp vs. Human DNA Sequences sheets, parts 1 and 2 (taped together)
- scissors

1. Comparing DNA

a. Compare the Chimp vs. Human DNA Sequences. The sequences are located on the X chromosome, and they are called Xq13.3. These are the small sections of the DNA that Svante Pääbo and his group use to make chimp/human comparisons. Look for any differences (mutations) between the chimp and human sequences.

b. How to read the chart:
- In the chart you will find the same stretch of DNA (about 2,700 nucleotides long) for a chimpanzee (top) and a human (bottom).
- The vertical lines show where the DNA is the same for chimps and humans.
- A tiny Pääbo figure shows where there's a difference in the chimp and human DNA.

c. How many Pääbos can you find?

Active Evolution Instruction

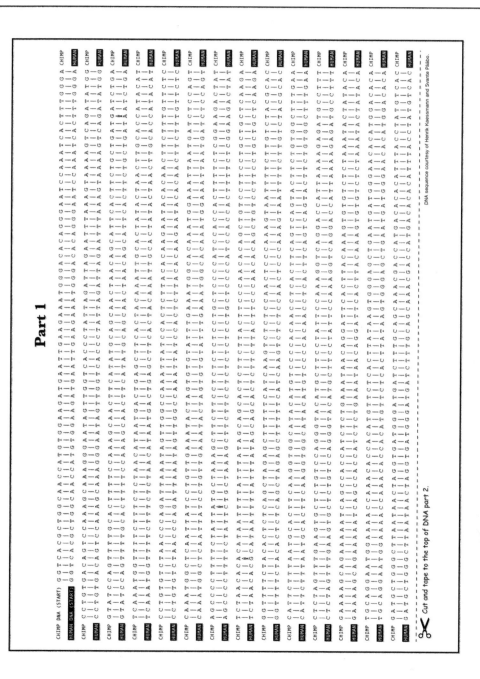

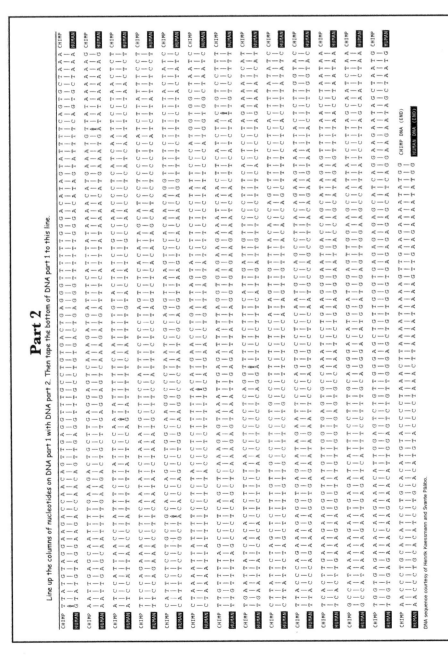

Active Evolution Instruction

More Resources
Internet

Staying up-to-date on topics such as evolution is a lot easier with the internet, but sometimes there is too much of a good thing. If you Google the word *evolution*, for example, you get more than 170,000,000 hits.

To narrow down the list with reliable, vetted sources, log into NSTA's SciLinks *(www.scilinks.org)* and use the codes provided throughout this book. In addition, here are a few favorite evolution websites from the SciLinks search team:

- Understanding Evolution *(http://evolution.berkeley.edu/evolibrary/home.php)* is a comprehensive site from the University of California Museum of Paleontology. The information provided on the site could be the basis for a complete course, study group, or self-study project on evolution. The teacher link in the right margin is a tremendous resource that directs readers to a searchable database of lesson ideas and to a link called "conceptual framework," which provides essential questions to help you organize a unit on evolution. You might start by clicking on the Evolution 101 *(http://evolution.berkeley.edu/evolibrary/article/evo_01)* link for a tutorial on evolution that would be appropriate for students as well.
- Evolution *(www.pbs.org/wgbh/evolution)* from PBS is another good source. Even if you don't have access to the video, the web-based resources are very good and visually stunning.
- NSTA has compiled a set of internet resources pertaining to evolution *(www.nsta.org/publications/evolution.aspx)*. NSTA provides a comprehensive list of links to the above websites, position papers on evolution, and additional resources and articles.
- Have you read any of Darwin's writings? Darwin's *On The Origin of Species (www.literature.org/authors/darwin-charles/the-origin-of-species)* is not an easy read, but it is the primary source for evolution. The website has links to his other works as well.
- In case the Galápagos Islands are not on your future travel itinerary, *Galápagos Education (www.nsta.org/publications/interactive/galapagos)* on the NSTA website is a great collection of background information, classroom investigations, and resources for teaching evolution.

- Teach Evolution and Make it Relevant (*www.evoled.org/default.htm*) pulls together resources from several websites into teaching units that may give you ideas on how to organize your instruction.
- Evolution and the Nature of Science Institutes (ENSI) (*www.indiana.edu/~ensiweb*), began in 1987 with the goal of improving the teaching of evolution in high school biology courses. ENSI encourages teachers to teach evolutionary thinking in the context of a more complete understanding of modern scientific thinking, with quality evolution instruction resources over the internet.
- *The TalkOrigins Archive* (*www.talkorigins.org*) is a collection of articles and essays that provide mainstream scientific responses to questions regarding the evolution/creation controversy, the origin of life, geology, biology, catastrophism, cosmology, and theology.

Books

For a broader look at topics in the evolution/creation controversy, consider curling up with one of the following books:

About the debate

- Alters, B. J., and S. M. Alters. 2001. *Defending evolution in the classroom: A guide to the creation/evolution controversy.* Sudbury, MA: Jones and Bartlett.
- Bybee, R. W. 2004. *Evolution in Perspective. The Science Teacher's Compendium.* Arlington, VA: NSTA Press.
- National Academy of Sciences and Institute of Medicine. 2008. *Science, evolution, and creationism.* Washington, DC: The National Academies Press.
- Scott, E. C. 2005. *Evolution vs. creationism: An introduction.* Berkeley: University of California Press.

About the science

- Carroll, S. B. 2005. *Endless forms most beautiful: The new science of evo devo and the making of the animal kingdom.* New York: Norton.
- Palumbi, S. R. 2001. *The evolution explosion: How humans cause rapid evolutionary change.* New York: Norton.

Active Evolution Instruction

About science and faith
- Collins, F. S. 2007. *The language of God: A scientist presents evidence for belief.* Waterville, ME: Wheeler.
- Miller, K. R. 1999. *Finding Darwin's God: A scientist's search for common ground between God and evolution.* New York: Cliff Street Books.

About classroom support
- BSCS. 2005. *The nature of science and the study of biological evolution.* Colorado Springs, CO: BSCS
- Diamond, J., et. al. 2006. *Virus and the whale: Exploring evolution in creatures small and large.* Arlington, VA: NSTA Press.
- Skehan, J. W., and C. Nelson. 2000. *The creation controversy and the science classroom.* Arlington, VA: NSTA Press.

Journals

NSTA journals also offer a number of relevant and insightful articles focusing on the topic of evolution. These articles are available online at *www.nsta.org/store.*

Science Scope
- Sandro, L., J. M. Constible, and R. E. Lee, Jr. 2007. Extreme arthropods. *Science Scope* 30(9): 24–32.

The Science Teacher
- Hess, L. 2006. Speak up! *The Science Teacher* 73(8): 10.
- Moore, M. 2004. State standards and evolution. *The Science Teacher* 71(6): 41–44.
- Parrott, A. N. 2005. From DNA to disorder. *The Science Teacher* 72(5): 34–39.
- Scalice, D., and K. Wilmoth. 2004. Astrobiology. *The Science Teacher* 71(10): 34–36.
- Scotchmoor, J., and A. Janulaw. 2005. Understanding evolution. *The Science Teacher* 72(9): 26–28.
- Schneegurt, M. A., A. N. Wedel, and E. W. Pokorski. 2004. Salty

microbiology. *The Science Teacher* 71(7): 40–43.
- Tieman, D., and G. Haxer. 2007. The discovery of *jelly bellicus*. *The Science Teacher* 74(2): 30–35.
- Walsh, J. 2007. Ring species through space and time—A class demo. *The Science Teacher* 74(9): 62–64.

Journal of College Science Teaching
- Benson, K. E. 2004. My brother's keeper: A case study in evolutionary biology and animal behavior. *Journal of College Science Teaching*. 34(2): 40–45.
- Gregg, T. 2007. Intelligent design: Jonathan Wells and the tree of life. *Journal of College Science Teaching* 36(6): 10–11.

Chapter 3: Social Challenges of Promoting Evolution-Only Instruction

Sometimes teachers face pressure from administrators, districts, parents, or students (or all of the above) to teach creation "science" or intelligent design theory, or even to present evidence against evolution. And sometimes teachers face pressure to just forget about teaching evolution altogether. What should you, as a teacher, do? Even though you understand evolution's important place in biology, geology, and astronomy, broaching the topic in the classroom may be difficult. As you read this chapter, which discusses issues that surround the teaching of evolution, reflect on your own teaching practices.

What's *Wrong* With "Teaching the Controversy"

Many teachers, parents, and even some administrators advocate "teaching the controversy" of creationism and evolution. According to them, evolution is "just one theory" among other "theories" that should be explored. Such instruction, they argue, would foster critical thinking: "It's only fair" to give creationism equal time with evolution, and when all the evidence is laid out, students will be better prepared to "make up their own minds." On the surface, this sounds reasonable. What, then, is the problem?

To begin with, within the scientific community, there is no controversy regarding the occurrence of evolution. Scientists do not debate whether living things descended from common ancestors. Evolution is a body of facts and observations that constitute the underpinning of the science of biology. Evolution is not a *theory* in layman's terms, but a *scientific theory*—the best explanation scientists have today for how life on Earth has changed.

Supporters of intelligent design claim their beliefs to be scientific theory as well; however, the concept of intelligent design, in contrast, is rooted in legal and religious, rather than scientific, debate. The movement originated in the 1980s with a group of creationists who were disappointed when "creation science" failed in the courts. In *McLean v. Arkansas Board of Education* a federal court ruled that creation science was not science but religion and thereby violated the First Amendment. Creation science also failed to win large-scale religious support because Catholics and mainstream Protestants did not believe in the biblical literalist theology on which creation science relied.

The founders of intelligent design sought to sidestep these problems by promoting a version of creationism that would satisfy First Amendment standards of church-state separation and still hold widespread appeal to Christians.

Social Challenges of Promoting Evolution-Only Instruction

They abandoned creation science's focus on a 10,000-year-old Earth, proof in geology of Noah's flood, and other claims of fact that had long been refuted. What evolved was a focus on the central idea of creationism, which is God's special creation of things in their present form, and especially on God's design of complex structures. Intelligent design holds that certain features of the universe and of living things are best explained by an intelligent cause, not an undirected process such as natural selection.

However, this notion of intelligent design, which gained increasing popularity in the mid-1990s and early 2000s among individuals associated with the Seattle-based Discovery Institute and which began making inroads into education through school board edicts, still did not meet the standard of scientific theory. The majority of the claims of intelligent design are not testable; hence, it is not science. The few claims that are testable, such as the supposed inability of evolution to produce "irreducibly complex" structures—an argument stating that certain biological systems, organs for example, are so complex that they must have been "intelligently designed," because if one component were removed, the system would not work—have been shown to be incorrect.

Scientists also have rejected intelligent design because the view of science it incorporates greatly differs from that of mainstream science. Intelligent design focuses on the claim that science cannot explain certain phenomena and therefore those phenomena must be explained by an "intelligent agent" (i.e., God). Normally in science something unexplained is not considered unexplain*able*. Rather such circumstances warrant further study.

Intelligent design eventually failed in the legal community as well. In the dramatic 2005 Federal District Court trial *Kitzmiller et al. v. Dover Area School District et al.*, a judge declared that it is unconstitutional to teach intelligent design because it is religion and not science. (See "Evolution Instruction in the Courtroom," p. 48, for more details.)

Studies do show that the number of teaching hours currently devoted to alternative theories like intelligent design is typically low. However that such instruction occurs at all in a science classroom wrongly conveys to students that these ideas should be accorded respect as scientific perspectives. The Discovery Institute has stopped advocating the inclusion of intelligent design in high school biology classes because, the group says, attempts to mandate the teaching of intelligent design only politicize the idea and hinder fair and open discussion of its merits within the scientific community. But many others have not—among them school board members, parents, and even some science teachers. Intelligent design should not be taught for both pedagogical and legal reasons. Because intelligent design is not science, or even incorrect science, it is pedagogically unsuitable for presentation in a science class. It is a matter of faith that should be discussed within a religious community, not advocated in any public school classroom.

> **Research shows . . .**
> - Data suggests that about one in seven biology teachers are creationist in orientation. Roughly one in six professed a "Young Earth" personal belief. About one in eight report that they teach creationism or intelligent design in a positive light (Berkman, Pacheco, and Plutzer 2008, Table S5).

Q: *Why not appease the anti-evolution folks and teach the "controversy" or the so-called "strengths and weaknesses" of evolution, which will clearly demonstrate why evolution is the most complete explanation about how life on Earth has changed and continues to change?*

A: This strategy was conceptualized by the leaders of the Discovery Institute's Center for Science and Culture as a ploy to avoid the theological implications of intelligent design and require teachers to discuss the evidence that supports and that which refutes the theory of evolution (Shaw 2005). This may be an attractive solution for those who wish to avoid confrontation with certain special interest groups. Yet while there might be a "political" controversy about evolution, there is no such thing within the scientific community. No evidence exists that refutes evolution, and the weaknesses often identified by proponents of "teach the controversy" tend to be "straw men" that can be easily blown away, earlier conclusions that eventually were rejected by scientists, misinterpretations of data, and unanswered questions that do not threaten the integrity of existing conclusions regarding biological evolution.

Q: *Evolution conflicts with my religion, which says God created the world as it appears today. Therefore, I cannot "believe" in evolution.*

A: Many kinds of questions can be asked. "Does God exist?" is a question of faith, while "What's the best movie currently showing?" is a question of opinion. "Should federal funds be used to support stem-cell research?" is a question of debate. "How are the genomes of

Social Challenges of Promoting Evolution-Only Instruction

the wolf and the dog the same?" is an empirical question. Empirical questions like this one are settled primarily by using evidence. "Has biological evolution occurred?" is an empirical question that has been answered by the scientific community through the accumulation of evidence over an extended period of time. "Has biological evolution occurred?" is not a question of faith where one's beliefs are the focus. The goal of science instruction is to teach science, not religious beliefs. A quality science education must include the theory of evolution, which shows that the universe has a history and that change through time has taken place. It is the basis of biology.

Evolution Instruction in the Courtroom

Legal decisions concerning creationism and evolution rest on the First Amendment to the U.S. Constitution. In part, it states, "Congress shall make no law respecting an establishment of religion, or prohibiting the free exercise thereof . . ." The Establishment and Free Exercise clauses taken together require that public institutions be religiously neutral: Schools can neither promote nor inhibit religious expression. Therefore, it is perfectly legal for you to teach about religion, though it must to be in a nondevotional context. You can describe a religion or religious views, but it is unconstitutional to say, "Buddha was right!" Similarly, you can discuss controversies involving religion, but it would be improper to take sides (such as "the Pilgrims were right to burn witches because witches are evil").

Let's look at what can and cannot *legally* be done in classroom instruction.

- A state/district/school *cannot* ban the teaching of evolution. The 1968 Supreme Court decision *Epperson v. Arkansas* struck down anti-evolution laws such as that under which John T. Scopes was tried in 1925 in Tennessee. As a result, an administrator cannot legally tell a teacher not to teach evolution.
- A state/district/school *cannot* require equal time for creationism or creation science. The Supreme Court in 1987 (*Edwards v. Aguillard*) struck down laws that would require "equal time" for evolution and creation science by noting that even if the word *science* was used, creation science really is religion in disguise, and therefore it is illegal to teach it.
- A state/district/school *can* require the teaching of evolution. Requiring a science teacher to teach evolution is simply and appro-

priately requiring the teacher to teach a scientific theory in biology class (*Peloza v. Capistrano Unified School District*).
- A teacher *cannot* teach creationism on his/her own. Some teachers teach creationism or creation science even though their district does not (and legally cannot) have a policy requiring it. Such "freelancing" is illegal (*Webster v. New Lenox School District*).
- A state/district/school *cannot* have a disclaimer that singles out evolution. Such a disclaimer that singles out evolution from all other scientific theories for special treatment (for example, as theory, not fact) has been declared unconstitutional by a Louisiana Federal District Court and its associated Appeals Court (*Freiler v. Tangipahoa Parish Board of Education*) and by a Georgia Federal District Court (*Selman et al. v. Cobb County School District et al.*).

Topic:
Evolution teaching resources

Go to:
www.SciLinks.org

Code: ETK001

As states have reviewed their science standards and districts have attempted to exercise more control over what happens in individual classrooms, evolution has been an issue in several well-publicized court cases. The most recent of these is *Kitzmiller et al. v. Dover Area School District et al.*, the 2005 case in which U.S. District Court Judge John E. Jones III struck down an intelligent design policy in Pennsylvania's Dover Area School District. The district's intelligent design policy included a statement in the science curriculum that "students will be made aware of gaps/problems in Darwin's Theory and other theories of evolution including, but not limited to, intelligent design." Teachers in the district were also required to announce to their biology classes that "intelligent design is an explanation of the origin of life that differs from Darwin's view. The reference book *Of Pandas and People* is available for students to see if they would like to explore this view in an effort to gain an understanding of what intelligent design actually involves. As is true with any theory, students are encouraged to keep an open mind." In his 139-page ruling, Judge Jones wrote, it was "abundantly clear that the Board's ID [intelligent design] Policy violates the Establishment Clause." Furthermore, Judge Jones ruled that "ID [intelligent design] cannot uncouple itself from its creationist, and thus religious, antecedents." In reference to whether intelligent design is science, Judge Jones wrote, it "is not science and cannot be adjudged a valid, accepted scientific theory as it has failed to publish in peer-reviewed journals, engage in research and testing, and gain acceptance in the scientific community." This was the first challenge to the constitutionality of teaching intelligent design in the public school science classroom.

Social Challenges of Promoting Evolution-Only Instruction

Earlier in 2005, in *Selman et al. v. Cobb County School District et al.*, U.S. District Court Judge Clarence Cooper ruled that an evolution warning label required in Cobb County textbooks violated the Establishment Clause of the First Amendment. The warning labels stated, "This textbook contains material on evolution. Evolution is a theory, not a fact, regarding the origin of living things. This material should be approached with an open mind, studied carefully, and critically considered." After the district court's decision, the stickers were removed from Cobb's textbooks. The school district, however, appealed to the 11th Circuit Court of Appeals and in May 2006 the Appeals Court remanded the case to the district court for clarification of the evidentiary record. On December 19, 2006, the lawsuit reached a settlement; the Cobb County School District agreed not to disclaim or denigrate evolution either orally or in written form.

Other cases over the past 40 years have been significant in their support for teaching evolution to the exclusion of creationism, creation science, or intelligent design in public school settings:

2000 Rodney LeVake v. Independent School District 656 et al.
High school biology teacher LeVake had argued for his right to teach "evidence both for and against the theory" of evolution. The school district concluded that his content did not match the curriculum, which required the teaching of evolution. The District Court ruled that LeVake did not have a free speech right to override the curriculum, nor was the district guilty of religious discrimination.

1997 Freiler v. Tangipahoa Parish Board of Education
A district court rejected a policy requiring teachers to read aloud a disclaimer whenever they taught about evolution, ostensibly to promote "critical thinking." Besides addressing disclaimer policies, the decision is noteworthy for recognizing that curriculum proposals for intelligent design are equivalent to proposals for teaching creation science.

1994 Peloza v. Capistrano School District
A Circuit Court of Appeals upheld a district court finding that a teacher's First Amendment right to free exercise of religion is not violated by a school district's requirement that evolution be taught in biology classes.

1990 Webster v. New Lenox School District
A Circuit Court of Appeals found that a school district may prohibit a

teacher from teaching creation science in fulfilling its responsibility to ensure that the First Amendment's Establishment Clause is not violated and that religious beliefs are not injected into the public school curriculum.

1987 Edwards v. Aguillard
The Supreme Court held unconstitutional Louisiana's "Creationism Act," which prohibited the teaching of evolution in public schools, except when it was accompanied by instruction in creation science.

1982 McLean v. Arkansas Board of Education
A federal court held that a statute requiring public schools to give balanced treatment to creation science and evolution science violated the Establishment Clause of the U.S. Constitution. The decision also gave a detailed definition of the term science and the court declared that creation science is not, in fact, a science.

1981 Segraves v. State of California
The court found that the California State Board of Education's Science Framework, as qualified by its antidogmatism policy, gave sufficient accommodation to the religious views of Segraves, contrary to his contention that class discussion of evolution prohibited his and his children's free exercise of religion.

1968 Epperson v. Arkansas
The Supreme Court invalidated an Arkansas statute that prohibited the teaching of evolution.

Evolution Instruction in the Classroom

Let's look at four case studies developed by the National Center for Science Education (NCSE). These case studies, based on real-life situations, show the challenges of teaching evolution in the public school setting from the perspectives of students, parents, teachers, and administrators. As you read, consider whether situations like these might occur in your community.

Social Challenges of Promoting Evolution-Only Instruction

Case 1: Strengths and weaknesses of modern evolutionary theory

The situation: The veteran teacher of an Advanced Placement (AP) Biology course asked students to read the textbook section on evolution and then asked them to think about the scientific interpretation of the evidence versus other ways of thinking. Using the fossil record as an example, the teacher explained to students that gaps in the record exist and that fossils represent the preservation of bone only. Therefore, the fossil record does not explain how extinct creatures lived or how they came into existence. Nor does the record show a progression, such as how gills became lungs. The teacher explained that because no complete record exists it cannot be assumed that these events happened.

The conundrum: One student shared the day's instruction with her parents, both of whom were knowledgeable about and interested in the natural sciences. They recognized that the mode of instruction did not include scientific evidence or support the current understanding of evolution as a foundational concept in biology. The girl's parents became very concerned about their daughter's preparation for the college-level science courses she would encounter in her potential premedicine major.

The parents' response: The parents first scheduled a conference with the teacher, requesting more details and verification from the teacher and specifically asking how she taught evolution and whether she taught about gaps in the fossil record. The parents asked the teacher where she had acquired her background information.

The teacher's response: The teacher verified that she was indeed teaching weaknesses in Darwin's theory of evolution and specifically that she was doing so to help students become critical thinkers.

The parents' concerns: The parents left the teacher conference still concerned about their daughter's level of understanding of this important biological concept and unsure of what further course of action should be taken. After research, they contacted NCSE for advice.

NCSE advice: NCSE suggested that the parents write a formal letter to the teacher explaining their position—that a strong science education includes evolution instruction without disclaimers—and that the parents plan to discuss their concerns with the science department chair. NCSE helped the parents prepare a statement for the department chair that included support for evolution instruction and information explaining why teaching "strengths and weaknesses" is not a valid pedagogical example of teaching critical thinking. In addition, NCSE put the parents in contact with a local university professor who offered to explain (to the student, her teacher, the school administrators, or whoever else needed convincing) the central role of evolution in the biological sciences as well as the expectations of prior knowledge in a college-level biology curriculum. NCSE also directed the student to the statement by The College Board on inclusion of evolution in the AP curriculum.

> **Research shows . . .**
> - Individuals' education level may have an impact on attitudes. Scientific knowledge is favorably correlated with evolution-only instruction. Only 23% of respondents correctly answered all questions designed to test general science knowledge (Coalition of Scientific Societies 2007, p. 6).

The department chair's response: The department chair empathized with the parents' concerns but minimized the problem, saying the class had finished their unit on evolution and that there was little reason to pursue the issue further. The teacher was well liked by many students and otherwise exemplary.

The parents' next steps: While this teacher's mode of evolution instruction seemed like a small, temporary problem, the parents were reluctant to let the issue go unresolved as they thought about the science education of future students. So although evolution instruction was no longer an issue in their daughter's class and seemed to be a minor part of the entire AP Biology curriculum, the parents decided to move up the chain of command, scheduling a conference with the school principal and district superintendent. The parents asked the university professor to accompany them to reinforce the importance of evolution to a strong foundation in biology and to explain that universities expect knowledge of evolution upon accepting AP credit. If the teacher refused to alter her approach, universities might stop accepting AP Biology credits from this particular high school.

The resolution: The teacher would teach evolution according to scientific and educational guidelines, and no longer use gaps in the fossil record to signify weakness of evolutionary theory. To rectify the potential harm already done, an evolutionary biologist would be invited to speak to the class.

Social Challenges of Promoting Evolution-Only Instruction

Q: *Why do some say that gaps exist in the fossil record, which indicates a weakness in the theory of evolution?*

A: It is important to recognize that fossils are only one source of evidence of the occurrence and history of evolution. Despite the consistency in the fossil record, some species do seem to appear abruptly. Scientists readily admit that the fossil record cannot adequately explain or document the complete history of life on Earth. However, when combined with biochemical, genetic, and structural evidence, the fossil record provides convincing evidence that life has evolved through time.

Case 2: The choice to opt-out of the evolution unit

The situation: At the beginning of the school year a few parents approached the ninth-grade biology teacher about teaching creationism as an alternative to evolution. The teacher advised the parents that his professional responsibility required teaching according to national and state standards, which clearly place evolution as a foundational concept in biology. When the teacher began his direct instructional unit on evolution, he received a note from the superintendent indicating that an old policy on the books allowed parents to have their students opt-out of any material they found offensive to their religious beliefs or values. Therefore, four students would be leaving class and the teacher was to provide alternative lessons for these students.

The teacher's response: The teacher, with the support of his department chair and the principal, called NCSE for advice.

NCSE advice: NCSE advised the teacher to obtain more information on the opt-out policy and determine if a similar state policy also exists. Next, the teacher should schedule a meeting with the superintendent requesting that the department chair and principal also be present. During the meeting, the teacher should

- clarify his position, restating that opting out of evolution creates a disruptive environment for the other students and that students who do not learn evolution will not be adequately prepared for future science courses and state exams; and

- request that the superintendent identify a pedagogically appropriate alternative assignment, noting that the law prohibited including creationism or intelligent design as alternative assignments.

NCSE also advised the teacher to show the following supporting documentation:

- A copy of the court decision in *Epperson v. Arkansas*, with key phrases highlighted: "There is and can be no doubt that the First Amendment does not permit the state to require that teaching and learning must be tailored to the principles or prohibitions of any religious sect or dogma."
- The NSTA and the National Association of Biology Teachers (NABT) statements on the importance of evolution instruction as a foundational concept in biology *(www.nsta.org/about/positions/evolution.aspx* and *http://nabt.org/sites/S1/index.php?p=65).*

The resolution: The teacher held an information night for parents, particularly parents who believed evolution conflicted with their religious beliefs. He started the session by emphasizing the nature of science, science process, and scientific thought. He showed the parents the evolution-creationism continuum, a graphical description of the many forms of evolutionists and creationists. He also brought in a religious member of the community who helped explain the difference between science and religion. In addition, the teacher provided references of scientists who have come forward about their religious positions (for example, Francis Collins and Kenneth Miller) and had NCSE's *Voices for Evolution* (Matsumura 1995), which can be accessed online at *www.ncseweb.org/article.asp?category=2,* and the AAAS's *The Evolution Dialogues* (Baker 2006) on hand. In the end, parents agreed that their children would remain in the class during the evolution section, if for no other reason than to be able to pass state exams. The teacher also worked with the superintendent to suggest new wording for the policy: "Recognizing the importance of learning in an uninterrupted and thorough manner, it shall be the policy of this School District that no student shall be excused from class on the grounds of personal or family objections to the material presented, except as required by state law."

Social Challenges of Promoting Evolution-Only Instruction

Case 3: Science curriculum policy and academic freedom

The situation: A parent attending a school board meeting happened to be present when the school board passed an academic freedom policy aimed at science education. The policy stated, "Since teaching some scientific subjects, such as biological evolution and global warming, can cause controversy, and some teachers are unsure how information on such subjects should be presented, the district shall assist teachers to find more effective ways to present these controversial parts of the science curriculum. Teachers shall help students understand, analyze, critique, and review the scientific strengths and weaknesses of existing scientific theories, thereby developing critical-thinking skills and appropriate responses to differences of opinion about controversial issues."

The conundrum: The policy sounded professional and geared toward helping teachers deal with difficult issues. However, a few flags went up for the parent when she considered it in more detail, especially since biological evolution was singled out as controversial and she knew that biological evolution is not controversial within the scientific community.

The parent's response: First she sought clarification from the school board, specifically asking how this policy would affect the curriculum and who would determine which topics were controversial. Would the school board, the teachers, or the students decide? The only topics listed (biological evolution and global warming) were not scientifically controversial, so was the school board defining controversial issues based on socially and/or politically controversial issues? If so, was social controversy an appropriate topic for science class, especially given all the other topics that teachers much teach?

The parent also spoke specifically with the teachers, asking what their plans were with respect to this proposal, and was careful to keep thorough notes on every conversation. A few teachers indicated that they were in favor of the new policy. One noted that Darwin himself questioned his own theory and that the teachers should be able to do the same, as a matter of fairness.

The parent's concern: After speaking with the school board and the teachers, the parent was convinced that the Science Curriculum Policy was designed to give teachers a way to introduce creationism or evidence against evolution. At this point, the parent contacted NCSE as well as other parents, asking if any of them were at all concerned about the implications of the new Science Curriculum Policy. In fact, a number of the parents were very concerned about the policy and they started working together to challenge the implementation of the policy.

NCSE advice: NCSE recommended that through transcripts, the parent evaluate whether any members of the board had training in the sciences and could therefore determine scientifically controversial topics. NCSE also helped the parents draft a letter to the school board, superintendent, principal, and department chair. The letter included an explanation of how the peer-review process of scientific research already provided the opportunity for proponents of creationism to persuade scientists of the scientific validity of their position, which they had been unable to do. The letter also included a clarification about the pedagogical inappropriateness of teaching about nonexistent controversies and emphasized that evolution was not controversial in the scientific community.

The letter also included a summary of the case of *Rodney LeVake v. Independent School District 656 et al.* As mentioned earlier in this chapter, LeVake had argued for his right to teach "evidence both for and against the theory" of evolution. The school district concluded that his content did not match the curriculum, which required the teaching of evolution. The district court ruled that LeVake did not have a free speech right to override the curriculum; nor was the district guilty of religious discrimination.

The resolution: Science won out. The teachers decided to continue teaching evolution as outlined in the state standards. The community rallied support around the issue as well and, in the next election, replaced the members of the school board who had voted in favor of the policy.

Case 4: Textbook adoption

The situation: It was science textbook adoption year and volunteer teachers were conducting in-depth reviews of eight different texts for possible use in regular and honors biology, completing detailed rubrics that would be used in the all-teacher vote. In a close decision, teachers chose *Biology* by Jones and Smith (pseudonyms) for their regular students. But the State Board of Education (SBOE) had other plans in mind. Although the Jones-Smith text was on the approved state list, the SBOE had recently placed its approval of the text "on hold" based on a review by a local university faculty member who chal-

Social Challenges of Promoting Evolution-Only Instruction

lenged the text's characterizations of Darwin's theory of evolution as being the foundation for all lessons about life, including survival of the fittest.

The department chair's response: The science department chairman had grave concerns about both the "hold" and the information on which the SBOE had based its decision. He first brought his concerns to his principal, who then shared them with the superintendent. The chair also notified his state's Citizens for Science group, which until this point had not heard about the "hold." (Citizens for Science is a network of grassroots organizations devoted to protecting and promoting science education.) The principal seemed to understand the chairman's position, but he was reluctant to take a stand. The superintendent, however, tried to help the department chair work with the SBOE in finding a compromise, one that would accommodate both parties.

The SBOE's response: First, the State Board suggested putting a disclaimer stating that "evolution is just a theory" in all Jones-Smith textbooks. The department chair quickly pointed out the outcome of *Selman et al. v. Cobb County School District et al.*, in which U.S. District Court Judge Clarence Cooper ruled that the evolution warning label required in Cobb County textbooks violated the Establishment Clause of the First Amendment.

The SBOE then suggested that supplementary materials be made available to students in addition to the Jones-Smith text and specifically recommended two resources: *Icons of Evolution: Science or Myth?* (Wells 2000), which describes how accepted evidence for evolution such as homologous structures and *Archaeopteryx* are not as supportive as science says, and *Explore Evolution* (Meyer et al. 2007), which claims to use an inquiry-based approach to teach students the case for and against the modern view of the theory of evolution.

The state's Citizens for Science response: The state's Citizens for Science organized its members and contacted the authors and publisher of the Jones-Smith text. Author Thomas Henry Jones wrote replies to the concerns on which the hold was based.

The resolution: Jones addressed the state Board of Education in person. Science teachers and college professors also supported the book's emphasis on evolution as sound science. They argued that high school biology students would receive an incomplete education and be ill-prepared for college course requirements if they did not learn such theories. After the discussion ended, the SBOE Chairman called for a vote of the board members. After the text received a simple majority of approvals, voting ceased and the hold was removed.

Local communities and states should have the universal power to decide what should be taught to their students. If a community wants to include evidence against evolution, shouldn't it be their prerogative?

 NSTA supports parental and community involvement in establishing the goals of science education and in securing the support needed to achieve them. Local communities and states should and currently do have a role in shaping science curricula. Currently, all states but Iowa have developed science standards that define with varied degrees of specificity what students should know and be able to do. These standards typically are approved through a political process that provides extended opportunity for public input. Evidence that reveals weaknesses in any given theory should be emphasized in the science curriculum as appropriate. The weaknesses, however, should be those recognized and subject to debate within the scientific community. Overall, science curriculum and instruction must be informed by and shaped by the knowledge gained through investigation and evidence, and at the same time directed toward achieving goals established through a collaborative process that involves educators, community participants, parents, and policy makers.

Leading the Way for Evolution Instruction

How can you feel confident in your evolution instruction methodology and perhaps even become a mentor for others? Take charge of the tone of your instruction, setting an example for your students and colleagues. Research shows that some messages and presentations are more effective than others, fostering an environment in which you have more time to focus on the instructional goals of your course and state testing needs.

More effective messages

Religious education is a private matter. School boards and government should not impose one religious viewpoint on public school students. Students should learn about religious ideas, such as creationism and intelligent design at

Social Challenges of Promoting Evolution-Only Instruction

church and at home and not in public school science classes. Treat student and parent religious beliefs with respect, but stand firm that evolution is science while creation science and intelligent design are not. Although studies show that those with fundamentalist religious beliefs are less accepting of evolutionary theory, it is not in conflict with understandings in most mainstream religions. The Jewish, Catholic, and most Protestant sects have no objection to evolutionary theory and many participants in those religions understand that when a school district tries to support creationism, it favors one religious view over another.

> **Research shows . . .**
> - About 63% of respondents from the general public ranked "developing medicines and curing diseases as the most important contribution of science to society," regardless of whether they were proponents of teaching evolution, creationism, or intelligent design (Coalition of Scientific Societies 2007, p. 6).

Evolution is the basis of modern medical science. Scientists are concerned that illnesses like avian flu may change and become harder to treat, causing millions of people to get sick or die. Evolution helps us understand this process and develop better medical responses. Rather than attacking the teaching of evolution, science students should be taught that evolution is critical to understanding medicine and the human body. Making connections between evolutionary biology and advancing other areas of medical research, such as understanding human gene function or the mechanisms by which antibiotic resistance develops, can nullify potential objections.

Less effective but still useful messages

Not all theories are the same. Scientific theories such as evolution describe the natural world and have been tested so often that scientists consider them to be reliable explanations. Other explanations such as creationism and intelligent design are not testable in the natural world. In public school science class, concentrate on teaching evolution because it is the only *scientific* explanation we have for the nature and history of life on Earth.

Evolution is universally accepted among mainstream scientists. While there may be some public debate about the details of evolution as an explanation for the nature and history of life on Earth, scientific acceptance of evolution is universal. Many creationism and intelligent design advocates would require schools to single out evolution for special criticism and would deny students a clear understanding of a scientific explanation that is as established and important as the existence of gravity.

How should I approach the topic of evolution to alleviate students' concerns or questions about it?

 You may not be able to alleviate students' concerns about how biological evolution fits into their mental models concerning the history and nature of life on Earth. Teachers should indicate that their responsibility is to plan and provide instruction that reflects the current knowledge and thought existing within the scientific community. As a result, teachers should strive to provide students with multiple opportunities to learn about current scientific thinking regarding biological evolution and the history of life on Earth. Because effective science instruction is characterized by inquiry and student questions, students' questions about biological evolution should be encouraged.

Chapter 4: What *You* Can Do

Now that you have read the *NSTA Tool Kit for Teaching Evolution*, you might ask yourself, "What can I do to advocate for evolution education?" If you are one of the teachers described in the Introduction who truncates evolution instruction or eliminates it entirely from your curriculum, we hope this guide has given you some ideas for overcoming the challenges of teaching evolution. If you are already an advocate for evolution in your instruction, we hope you will press on and find ways to help your colleagues.

Research shows . . .
- Data suggest that having all teachers complete a course in evolutionary biology would have a substantial impact on the emphasis on evolution and its centrality in high school biology courses (Berkman, Pacheco, and Plutzer 2008, p. 4).

Community Awareness

Although teaching is in itself a full day, make time to stay abreast of community attitudes with regard to evolution instruction. Individual complaints can quickly morph into a full-fledged movement. At the writing of this book, for example, Florida's recently adopted science standards, which include explicitly stated standards on evolution, are under fire. Although the standards were drafted only after much careful research and consideration had been given to the nature of science, a bill that would give teachers the "academic freedom" to include nonscientific content, has been proposed to the state legislature in response to individual outcry.

If your community faces a similar situation, step in and provide science background for parents and administrators. You may also want to have discussions with your colleagues and friends, and send letters to your legislators to maintain the integrity of your science instruction.

The NSTA Community

NSTA invites all teachers to share their best practices, challenges, ideas, and tips for motivating students while emphasizing the importance of evolution to the study of biology. Just sign on to *www.nsta.org/evolution* and please share how you encouraged colleagues, as well as parents and the community, to understand that the science teacher must be focused on science, not religion. After all, *you* are the one preparing students for a future based in science and technology.

What *You* Can Do

Need More Help?

In conclusion, please consider one last point: A 1999 study by two Oklahoma science educators revealed that "membership in professional organizations" was a strong factor in emphasizing evolution instruction (Weld and McNew 2004). No wonder. Supporting you as a professional, especially where potentially controversial topics are concerned, is a primary objective of not only NSTA but also many other organizations. Each organization listed below offers support if conflicts arise in your evolution instruction endeavors. If you are not already a member, consider joining one or more of these organizations. Each offers print and online resources you can tap as well as hundreds of collegial footprints into which you may step. You aren't alone, and you aren't the first to deal with this issue.

National Science Teachers Association (NSTA) (*www.nsta.org*)

NSTA strongly supports the position that evolution is a major unifying concept in science and should be included in the K–12 science education frameworks and curricula. The NSTA's position statement on the teaching of evolution, science background for this position, SciLinks, and many other resources can be found on its website.

National Center for Science Education (NCSE) (*www.natcenscied.org*)

NCSE defends the teaching of evolution in public schools. It is a nationally recognized clearinghouse for information and advice to keep evolution in the science classroom and creationism out. NCSE is the only national organization specializing in this issue.

American Association for the Advancement of Science (AAAS) (*www.aaas.org*)

AAAS is an international nonprofit organization dedicated to advancing science around the world by serving as educator, leader, spokesperson, and professional association. AAAS played a prominent role in responding to efforts in Kansas, Pennsylvania, and elsewhere to weaken or compromise the teaching of evolution in public school science classrooms. Background from the AAAS special event *Evolution on the Front Line*, held in 2006, along with other resources to support you in teaching evolutionary theory, may be found on the AAAS website.

National Association of Biology Teachers (NABT) *(www.nabt.org)*

NABT empowers educators to provide the best possible biology and life science education for all students. Its position statement on the teaching of evolution begins: "As stated in *The American Biology Teacher* by the eminent scientist Theodosius Dobzhansky (1973), 'Nothing in biology makes sense except in the light of evolution.'" Read the rest of their statement online at *http://nabt.org/sites/S1/index.php?p=65*. Two other organizations have endorsed NABT's position on teaching evolution: The Society for the Study of Evolution *(www.evolutionsociety.org)* and The American Association of Physical Anthropologists *(http://physanth.org)*.

The NABT site also offers additional resources, background materials, and support concerning the topic of evolution.

National Council for the Social Studies (NCSS) *(www.socialstudies.org)*

NCSS recognizes that recent court decisions finding the teaching of creationism and intelligent design in the public school science curriculum to be unconstitutional fueled efforts to introduce these beliefs into the social studies curriculum. Although NCSS believes in the open and thoughtful discussion of ideas, public school classrooms are not the place for the teaching of religious beliefs. Social studies is the forum for open analysis and discussion of historical, social, economic, geographic, political, and global issues. NCSS's recommendations seek to include the study of intelligent design within that framework. Read the rest of their position statement online at *www.socialstudies.org/positions/intelligentdesign*.

The National Academies *(www.nationalacademies.org)*

The National Academies, created by congressional charter in 1863, comprises four organizations: the National Academy of Sciences, the National Academy of Engineering, the Institute of Medicine, and the National Research Council. They produce books, position statements, and

What *You* Can Do

additional resources on evolution education and research, partnering with other organizations such as NSTA to strengthen their position. Materials such as their completely updated edition of the landmark booklet *Science and Creationism* can be found online at *www.nap.edu/html/creationism/*.

Smithsonian Institution (*www.si.edu*)

The Smithsonian Institution is uniquely different from a government agency or a university in that federal appropriations provide the core support for its science efforts, museum functions, and infrastructure, while trust resources, including external grants and private donations, provide the support for new discoveries and investigations and the global exchange of new ideas. The Institution's five-year plan includes a focus on understanding the formation and evolution of the Earth and similar planets. Visit the Smithsonian in person or online for resources and to find out more about their mission.

References

Baker, C. 2006. *The evolution dialogues: Science, christianity, and the quest for understanding.* Washington, DC: AAAS.

Berkman, M. B., J. S. Pacheco, and E. Plutzer. 2008. Evolution and creationism in America's classrooms: A national portrait. *PLoS Biology* 6(5): 920–924.

Bybee, R. 1997. *Achieving scientific literacy: From purpose to practice.* Portsmouth, NH: Heinemann.

Coalition of Scientific Societies. 2007. You say you want an evolution? A role for scientists in science education. *http://evolution.faseb.org/sciencecoalition.*

Darwin, C. 1959. *The autobiography of Charles Darwin: With original omissions restored.* New York: Norton.

Diamond, J., ed. 2006. *Virus and the whale: Exploring evolution in creatures small and large.* Arlington, VA: NSTA Press.

Matsumura, M. 1995. *Voices for evolution.* Oakland, CA: NCSE.

Meyer, S. C., S. Minnich, et al. 2007. *Explore evolution.* Melbourne and London: Hill House Publishers Ltd.

National Academy of Sciences (NAS). 1998. *Teaching about evolution and the nature of science.* Washington, DC: National Academy Press.

National Academy of Sciences and Institute of Medicine. 2008. *Science, evolution, and creationism.* Washington, DC: The National Academies Press.

National Research Council (NRC). 1996. *National science education standards.* Washington, DC: National Academy Press.

Omenn, G.S. 2006. Welcome and teacher survey. Address given at the AAAS evolution on the front line: An event for St. Louis–area teachers, St. Louis. *www.aaas.org/programs/centers/pe/evoline.*

Pew Forum on Religion and Public Life. 2005. Public divided on origins of life. *http://pewforum.org/surveys/origins.*

Schneegurt, M. A., A. N. Wedel, and E. W. Pokorski. 2004. Salty microbiology. *The Science Teacher* 71(7): 40–43.

Shaw, L. *Seattle Times.* 2005. Theory of Intelligent Design: A Debate Evolves. March 31.

Skehan, J. W., and C. Nelson. 2000. *The creation controversy & the science classroom.* Arlington, VA: NSTA Press.

Texley, J., and T. Kwan. 2007. MRSA and the science classroom. *NSTA Reports* 19(4): 1, 4–5.

UCMP Berkeley. Jean-Baptiste Lamarck. *www.ucmp.berkeley.edu/history/lamarck.html*

References

Weld, J., and J. C. McNew. 2004. *Evolution in perspective: The science teacher's compendium.* Arlington, VA: NSTA Press.

Wells, J. 2000. *Icons of evolution: Science or myth? Why much of what we teach about evolution is wrong.* Washington, DC: Regnery Publishing.

Index

*Page numbers in **boldface** type refer to figures or tables.*

A

AAAS (American Association for the Advancement of Science), ix, 55, 64–65
Academic freedom and science curriculum policy, 56–57
American Association for the Advancement of Science (AAAS), ix, 55, 64–65
Ampicillin resistance, **29**
Anatomy, comparative, 12–13, 18
Antibacterial soaps, 25–27
Antibiotic resistance, 25–30
   developing a public service campaign about, 30
   drug cost and, 27
   halotolerant bacteria for study of, 28–29
   internet research on, 30
   prevention of, 27
   resources about, 27
   to specific drugs, **29**
   student activities on, 28
Anti-evolution groups, xiii, **7**
Anti-evolution laws, **6,** 48. *See also* Legal decisions

B

Bacteria, 26
   antibiotic-resistant, 26–30, **29**
   culturing of, 28
   evolution of, 26
   generation times of pathogens, **29**
   halotolerant, 28–29
   horizontal gene transfer between, 26
Biblical flood, **7,** 8, 46
Biogeography, 13
Bioinformatics, 3
Bryan, William Jennings, **6**
BSCS 5E Instruction Model, 18
Buckland, William, 8

C

Catastrophism, 5, 8
Centers for Disease Control and Prevention, 27
Chromosomes, 14
Church-state separation, 45, 48
Cladogram(s), 18–25
   assessment of, 24–25
   characteristics of organisms selected for, **19**
   definition of, 19
   development of, 19–21, **20**
   for Hawaiian fruit flies, 23–24, **25**
   interpreting branches of, **21,** 21–22
   preparing character cards for, **22,** 22–23
   for primates, 31, **32**
   relative passage of geologic time on, 21
   sample of, **24**
   student construction of, 23
Coalition of Scientific Societies, xiv
Collins, Francis, 55
Community awareness, 63
Community role in shaping science curriculum policy, 59
Comparative anatomy, 12–13
   showing evolutionary relationships with cladograms, 18–25
Comparative genomics, 3, 11–12

Controversy about evolution instruction, ix, xiii–xiv, 17, 45–47, 56
Creation Controversy & The Science Classroom, The, 17
Creationism, xiii, 4, **6, 7**. *See also* Intelligent design
   Americans in favor of teaching, xiv
   Biblical flood and, **7,** 8, 46
   instruction outside of public schools, 59–60
   legal decisions on teaching of, 46, 48–51
   percentage of biology teachers providing instruction on, 47
   science curriculum policy and teaching of, 56–57
   teaching controversy of evolution and, 45–47
   Young Earth creationists, **7,** 47
Crick, Francis, **6,** 11
Cuvier, Georges, **4,** 5
Cytochrome c, 13

D
Darwin, Charles, 3, **5,** 6, 8–11, 40
Deoxyribonucleic acid (DNA), **6,** 11
   DNA comparisons to show evolutionary relationships, 32—39
   evaluating student understanding of, 33
Developmental biology, 12, 13
Discovery Institute, 46, 47
DNA. *See* Deoxyribonucleic acid
Dobzhansky, Theodosius, 11

E
Edwards v. Aguillard, **7,** 48, 51
ENSI (Evolution and the Nature of Science Institutes), 41
Epperson v. Arkansas, **7,** 48, 51, 55
Erythromycin resistance, **29**
Escherichia coli, 27, **29**
Essay on the Principles of Population, An, 10
Everson v. Board of Education, **6**
Evo-devo (evolutionary developmental biology), 12
Evolution
   biology teachers' beliefs about, 3
   Darwin's thought processes on, 8–10
   debates about "how" of, 3, 11
   definition of, 2–3
   evidence for, 12–13
   genetics of, 14–15
   history of evolutionary thinking, 3–8, **4–7**
   "proof" of, 2
   so-called "weakness" of, xiii, 47
Evolution (website), 40
Evolution and the Nature of Science Institutes (ENSI), 41
Evolution Dialogues, The, 55
Evolution instruction, 17–41
   activities for, 18–41
      DNA comparisons to show evolutionary relationships, 33, 34–39
      natural selection and antibiotic resistance, 25–30
      ring species to demonstrate human evolution, 30–33
      showing evolutionary relationships with cladograms, 18–25
   alleviating students' concerns about, 61
   attitudes of Americans about, xiii
   average time teachers spend on, xi–xii, xiv, 17
   BSCS 5E Instruction Model and, 18
   case studies on, 51–59
      choice to opt-out of evolution unit, 54–55
      science curriculum policy and academic freedom, 56–57
      strengths and weaknesses of modern evolutionary theory, 52–54
      textbook adoption, 57–59
   community awareness about, 63
   as conflict with teacher's religious beliefs, 47–48
   controversy about, ix, xiii–xiv, 17, 45–47
   frustration about dialogue about, xiii
   impact of teachers' personal beliefs on, xiv
   leading the way for, 59–61
   legal decisions on, 46, 48–51
   National Science Education Standards pertaining to, **x–xi**
   percentage of biology teachers providing, xi
   professional organizations supporting, 64–66

question-and-answer document for, xii–xiii
reasons for avoiding, ix
resources for, 40–43
social challenges of, xiii, 45–61
Evolutionary developmental biology (evo-devo), 12
Explore Evolution, 58
Extinct organisms, 5

F
First Amendment of U.S. Constitution, 48–50, 55. *See also* Legal decisions
Fisher, Ronald, 11
Fossil record, 5, 12, 54
Freiler v. Tangipahoa Parish Board of Education, 49, 50

G
Galápagos Education, 40
Galápagos Islands, 3, 8, **9,** 40
Genesis Flood, The, **7**
Genetics, 3, **7,** 11–15
  alleles of genes, 11, 14–15
  bacterial horizontal gene transfer, 26
  chromosomes, 14
  comparative genomics, 3, 11–12
  conceptualization of DNA, **6,** 11
  genetic drift, 15
  genetic recombination, 13
  genetic variation, 11–12, 14–15
  Human Genome Project, **7,** 11
  molecular, 13
  natural selection, 10–11, 14
  of populations, 11, 14–15
  speciation, 11, 15
  tool kit genes, 12–13
Gould, Stephen Jay, 12

H
Histoire naturelle, 4, **44**
HMS *Beagle*, 8, **9,** 11
Human Genome Project, **7,** 11
Hutton, James, **4,** 4–5, 8

I
Icons of Evolution: Science or Myth?, 58
Inheritance, 6, 11. *See also* Genetics
Institute for Creation Research, **7**
Institute of Medicine, 3, 65

Intelligent design, ix, **7,** 45–46. *See also* Creationism
  attempts to mandate teaching of, 46
  instruction outside of public schools, 59–60
  legal decisions on teaching of, 46, 48–51
  percentage of biology teachers providing instruction on, 47
  reasons for not teaching, 47
  scientists' rejection of, 46
  time spent by biology teachers on teaching, xiv, 46
Internet resources, 40–41. *See also* SciLinks
  professional organizations, 64–66

J
Journal of College Science Teaching, 43

K
Kaessmann, Henrik, 32
Kitzmiller et al. v. Dover Area School District et al., **7,** 46, 49

L
Lamarck, Jean-Baptiste, **4,** 5–6
Leclerc, George-Louis, 4, **4**
Legal decisions, 46, 48–51
  Edwards v. Aguillard, **7,** 48, 51
  Epperson v. Arkansas, **7,** 48, 51, 55
  Everson v. Board of Education, **6**
  Freiler v. Tangipahoa Parish Board of Education, 49, 50
  Kitzmiller et al. v. Dover Area School District et al., **7,** 46, 49
  McLean v. Arkansas Board of Education, **7,** 45, 51
  Peloza v. Capistrano Unified School District, 49, 50
  Rodney LeVake v. Independent School District 656 et al., 50, 57
  Segreaves v. State of California, 51
  Selman et al. v. Cobb County School District et al., 49, 50, 58
  Webster v. New Lenox School District, 49, 50–51
Lyell, Charles, **5,** 8

M
Malthus, Thomas, 10

Max Planck Institute for Evolutionary Anthropology, 32
Mayo Clinic, 27
McLean v. Arkansas Board of Education, **7**, 45, 51
Mendel, Gregor, 11
Methicillin resistance, **29**
Miller, Kenneth, 55
Molecular clock, 34
Molecular genetics, 13
Morris, Henry, **7**
Multidrug-resistant infections, 27
Mutations, 11, 14
   antibiotic resistance and, 27
   DNA comparisons to show evolutionary relationships, 34–39
   molecular clock and, 34
Mycobacterium tuberculosis, **29**

N

NABT (National Association of Biology Teachers), 55, 65
National Academies, The, 65–66
National Academy of Engineering, 65
National Academy of Sciences, 3, 65
National Association of Biology Teachers (NABT), 55, 65
National Center for Science Education (NCSE), x, 51–57, 64
National Council for the Social Studies (NCSS), 65
National Institutes of Health, 30
National Research Council, 65
National Science Education Standards, **x–xi**
National Science Teachers Association (NSTA), x, xii, xiii, xv, 40, 55, 59, 63–64, 66
Natural selection, 10–11, 14
   antibiotic resistance and, 25–30
   vs. intelligent design, 46
NCSE (National Center for Science Education), x, 51–57, 64
NCSS (National Council for the Social Studies), 65
Nelson, Craig, 17
Neo-Darwinism, **6**
Northwest Regional Educational Laboratory's 6+1 Trait Assessment Scoring Guide, 24–25
NSTA (National Science Teachers Association), x, xii, xiii, xv, 40, 55, 59, 63–64, 66
NSTA Reports, 28

O

Of Pandas and People, 49
"Old" Earth concept, 5
On the Origin of Species, 40
Opting-out of evolution unit, 54–55

P

Pääbo, Svante, 32, 34, 37
Paired box gene 6 (Pax6), 12–13
Paley, William, **4**
Peloza v. Capistrano Unified School District, 49, 50
Penicillin resistance, **29**
Pew Forum on Religion and Public Life, xiv
Phylogenetic trees, 19. *See also* Cladogram(s)
Phylogeny, 19
Price, George McCready, **6**
Primate evolutionary tree, 31–33, **32**
Professional organizations, 64–66
Public's view of goals of science education, 17

R

Religious beliefs, xiii, xiv, 1, **6**, **7**, 45–48, 55, 59–60. *See also* Creationism; Intelligent design
Ring species, 30–33
   definition of, 30
   distribution of, 31
   DNA comparisons to show evolutionary relationships, 32–39
   examples of, 31
   founder species for, 31
   primate evolutionary tree, 31–33, **32**
Rodney LeVake v. Independent School District 656 et al., 50, 57

S

Sanger, Frederick, **7**
SAT Reasoning Test Essay Scoring Guide, 24
Science, 1
Science, Evolution, and Creationism, 3

Science curriculum policy
   academic freedom and, 56–57
   community role in shaping of, 59
Science Scope, 42
Science Teacher, The, 29, 42–43
Science textbook adoption, 57–59
Scientific explanations, 1
Scientific literacy, 17
Scientific methodology, 1
Scientific theories, 1–2, 17, 45, 60
SciLinks, xii, 40
   antibiotic resistance, 26
   Darwin, 10
   evolution, 3
   evolution teaching resources, x, 49
   genome research, 12
   history of evolution, 8
   human evolution, 30
   phylogenetic trees, 21
   speciation, 14
   species/speciation, 32
Scopes Trial, **6,** 48
Segreaves v. State of California, 51
Selman et al. v. Cobb County School District et al., 49, 50, 58
Sickle-cell trait, 14
Skoog, Gerald, xii
Smithsonian Institution, 66
Social challenges to evolution-only instruction, xiii, 45–61
   case studies on, 51–59
      choice to opt-out of evolution unit, 54–55
      science curriculum policy and academic freedom, 56–57
      strengths and weaknesses of modern evolutionary theory, 52–54
      textbook adoption, 57–59
   legal decisions, 48–51
   what's wrong with "teaching the controversy?", 45–48
Speciation, 11, 15
   ring species, 30–33

St. Hilaire, Étienne Geoffroy, **5,** 8
Staphylococcus, 28, **29**
Streptococcus, 28
Streptomycin resistance, **29**

T
TalkOrigins Archive, The, 41
Teach Evolution and Make it Relevant (website), 40–41
Tetracycline resistance, **29**
Textbook adoption, 57–59
Theistic evolutionists, **5**
Theories, scientific, 1–2, 17
Theory of evolution, 2, 45
   addressing concerns about strengths and weaknesses of, 52–54
   Darwin's thought processes, 8–10
   modern synthesis of, 10–13
Timeline of evolutionary thinking, 3–8, **4–7**
Tool kit genes, 12–13
Treponema pallidum, **29**

U
Understanding Evolution (website), 40
Uniformitarianism, 5, 8
Ussher, James, 4, **4**

V
Venter, Craig, 12
Virus and the Whale, 33
Voices for Evolution, 55

W
Wallace, Alfred Russell, **5,** 10
Watson, James, **6,** 11
Webster v. New Lenox School District, 49, 50–51
Whitcomb, John, **7**
World Health Organization, 27
Wright, Sewell, 11

Y
Young Earth creationists, **7,** 47